APPLIED STRESS ANALYSIS

ELLIS HORWOOD SERIES IN MECHANICAL ENGINEERING

Series Editor: J.M. ALEXANDER, Professor of Mechanical Engineering, University of Surrey, and Stocker Visiting Professor of Engineering and Technology, Ohio University

The series has two objectives: of satisfying the requirements of post-graduate and mid-career engineers, and of providing clear and modern texts for more basic undergraduate topics. It is also the intention to include English translation of outstanding texts from other languages, introducing works of international merit. Ideas for enlarging the series are always welcomed.

Alexander, J.M.	Strength of Materials: Vol 1: Fundamentals; Vol. 2: Applications
Alexander, J.M. & Rowe, G.	Engineering Materials in Manufacturing Technology*
Alexander, J.M. & Brewer, R.C.	Technology of Engineering Manufacture
Atkins A.G. & Mai, Y.W.	Elastic and Plastic Fracture
Beards, C.	Vibration Analysis and Control System Dynamics
Beards C.	Structural Vibration Analysis
Besant, C.B. & C.W.K. Lui	Computer-aided Design and Manufacture, 3rd Edition
Borkowski, J. and Szymanski, A.	Technology of Abrasives and Abrasive Tools*
Borkowski, J. and Szymanski, A.	Uses of Abrasives and Abrasive Tools*
Brook, R. and Howard, I.C.	Introductory Fracture Mechanics*
Cameron, A.	Basic Lubrication Theory, 3rd Edition
Collar, A.R. & Simpson, A.	Matrices and Engineering Dynamics
Dowling, A.P. & Ffowcs-Williams, J. E.	Sound and Sources of Sound
Edmunds, H.G.	Mechanical Foundations of Engineering Science
Fenner, D.N.	Engineering Stress Analysis*
Fenner, R.T.	Engineering Elasticity
Ford, Sir Hugh, FRS & Alexander, J.M.	Advanced Mechanics of Materials, 2nd Edition
Gallagher, C.C. & Knight, W.A.	Group Technology Production Methods in Manufacture
Gohar, R.	Elastohydrodynamics
Haddad, S.D. & Watson, N.	Principles and Performance in Diesel Engineering
Haddad, S.D. & Watson, N.	Design and Applications in Diesel Engineering
Irons, B.M. & Ahmad, S.	Techniques of Finite Elements
Irons, B.M. & Shrive, N.	Finite Element Primer
Johnson, W. & Mellor, P.B.	Engineering Plasticity
Juhasz, A.Z. and Opoczky, L.	Mechanical Activation of Silicates by Grinding*
Kleiber, M.	Incremental Finite Element Models in Non-linear Solid Mechanics*
Launder, A.D., Gosman, B.E. & Reece, G.	Computer-aided Engineering: Heat Transfer and Fluid Flow
Leech, D.J. & Turner, B.T.	Engineering Design for Profit
McCloy, D. & Martin, H.R.	Control of Fluid Power: Analysis and Design 2nd (Revised) Edition
Osyczka, A.	Multicriterion Optimisation in Engineering
Oxley, P.	Mechanics of Machining*
Piszcek, K. and Niziol, J.	Random Vibration of Mechanical Systems
Prentis, J.M.	Dynamics of Mechanical Systems, 2nd Editon
Renton, J.D.	Applied Engineering Elasticity*
Richards, T.H.	Energy Methods in Vibration Analysis
Richards, T.H.	Energy Methods in Stress Analysis: with Introduction to Finite Element Techniques
Ross, C.T.F.	Computational Methods in Structural and Continuum Mechanics
Ross, C.T.F.	Finite Element Programs for Axisymmetric Problems in Engineering
Ross, C.T.F.	Finite Element Methods in Structural Mechanics
Ross, C.T.F.	Applied Stress Analysis
Sawczuk, A.	Mechanics of Plastic Structures
Sherwin, K.	Engineering Design for Performance
Szczepinski, W. & Szlagowski, J.	Plastic Design of Complex Shape Structured Elements*
Thring, M.W.	Robots and Telechirs
Walshaw, A.C.	Mechanical Vibrations with Applications
Williams, J.G.	Fracture Mechanics of Polymers
Williams, J.G.	Stress Analysis of Polymers 2nd (Revised) Edition

*In preparation

APPLIED STRESS ANALYSIS

C. T. F. ROSS, B.Sc.(Hons.), Ph.D.
Reader in Applied Mechanics
Portsmouth Polytechnic

ELLIS HORWOOD LIMITED
Publishers · Chichester

Halsted Press: a division of
JOHN WILEY & SONS
New York · Chichester · Brisbane · Toronto

First published in 1987 by
ELLIS HORWOOD LIMITED
Market Cross House, Cooper Street,
Chichester, West Sussex, PO19 1EB, England
The publisher's colophon is reproduced from James Gillison's drawing of the ancient Market Cross, Chichester.

Distributors:

Australia and New Zealand:
JACARANDA WILEY LIMITED
GPO Box 859, Brisbane, Queensland 4001, Australia

Canada:
JOHN WILEY & SONS CANADA LIMITED
22 Worcester Road, Rexdale, Ontario, Canada

Europe and Africa:
JOHN WILEY & SONS LIMITED
Baffins Lane, Chichester, West Sussex, England

North and South America and the rest of the world:
Halsted Press: a division of
JOHN WILEY & SONS
605 Third Avenue, New York, NY 10158, USA

British Library Cataloguing in Publication Data
Ross, C.T.F.
Applied stress analysis. —
(Ellis Horwood series in mechanical engineering)
1. Strains and stresses
I. Title
620.1'123   TA407

Library of Congress Card No. 86–24182

ISBN 0–7458–0077–7 (Ellis Horwood Limited — Library Edn.)
ISBN 0–7458–0155–2 (Ellis Horwood — Student Edn.)
ISBN 0–470–20767–1 (Halsted Press)

Phototypeset in Times by Ellis Horwood Limited
Printed in Great Britian by R.J. Acford, Chichester

Table of Contents

10   **Contents**

Author's preface

Failure of most components, whether they be electrical or mechanical devices, can be attributed to stress values. These components vary from small electrical switches to supertankers, or from automobile components to aircraft structures or bridges.

Thus, some knowledge of stress analysis is of much importance to all engineers, whether they be electrical engineers or mechanical engineers, or civil engineers or naval architects, or others.

Although, today, there is much emphasis on designing structures by methods dependent on computers, it is very necessary for the engineer to be capable of interpreting his/her computer output, and also that he/she knows how to use the computer program correctly.

The importance of this book, therefore, is that it introduces the fundamental concepts and principles of statics and stress analysis, and then it applies these concepts and principles to a large number of practical problems which do not require computers. The present author considers that it is essential for all engineers who are involved in structural design, whether they use computer methods or not, to be at least familiar with the fundmental concepts and principles that are discussed and demonstrated in this book.

This book should appeal to undergraduates in all branches of engineering construction and architecture and also to other students on B.T.E.C. and similar courses in engineering and construction.

The book contains a large number of worked examples, which are worked out in much detail, so that many readers will be able to grasp these concepts by working under their own initiative. Most of the chapters contain a section on "Examples for Practice", where the reader can test his/her newly acquired skills.

Chapter 1 is on "Statics", and after an introduction is made on the principles of statics, the method is applied to plane pin-jointed trusses, and to the calculation of bending moments and shearing forces in beams. The stress analysis of cables, supporting distributed and point loads, is also investigated.

Chapter 2 is on "Simple Stress and Strain", and after some fundamental definitions are made, application is made to a number of practical problems, including compound bars and problems involving stresses induced by temperature change.

Chapter 3 is on "Geometrical Properties", and shows the reader how to calculate second moments of area and the positions of centroids for a number of different two-dimensional shapes. These principles are then extended to "built-up" sections, such as the cross-sections of "I" beams, tees, etc.

Chapter 4 is on "Stresses due to Bending of Symmetrical Sections", and after proving the well-known formulae for bending stresses, the method is applied to a number of practical problems involving bending stresses, and also the problems on combined bending and direct stress.

Chapter 5 is on "Beam Deflections due to Bending", and after deriving the differential equation relating deflection and bending moment, based on small-deflection elastic theory, the equation is applied to a number of statically determinate and statically indeterminate beams. The area-moment theorem is also derived, and then it is applied to the deflection of a cantilever with a varying cross-section.

Chapter 6 is on the "Torsion of Circular Sections", and after deriving the well-known torsional formulae, these formulae are applied to a number of circular section shafts, including compound shafts. The method is also applied to the stress analysis of close-coiled helical springs.

Chapter 7 is on "Complex Stress and Strain", and commences with deriving the equations for direct stress and shear stress at any angle to the co-ordinate stresses. These relationships are later extended to determine the principal stresses in terms of the co-ordinate stresses. The method is also applied to two-dimensional strains, and the equations for both two-dimensional stress and two-dimensional strain are applied to a number of practical problems, including the analysis of strains recorded from shear pairs and strain rosettes. Mohr's circle for stress and strain is also introduced.

Chapter 8 is on "Membrane Theory for Thin-walled Circular Cylinders and Spheres", and commences with deriving the elementary formulae for hoop and longitudinal stress in a thin-walled circular cylinder under uniform internal pressure.

A similar process is also used for determining the membrane stress in a thin-walled spherical shell under uniform internal pressure. Application of these formulae is then made to a number of practical examples.

Chapter 9 is on "Energy Methods", and commences by stating the principles of the most popular energy theorems in stress analysis. Expressions are then derived for the strain energy of rods, beams and torque bars, and applications of these expressions are then made to a number of problems involving thin curved bars and rigid-jointed frames.

The method is also used for investigating stresses and bending moments in rods and beams under impact, and an application is also made to a beam supported by a wire.

The unit load method is also introduced.

Chapter 10 is on "Experimental Strain Analysis", and it gives an exposure of a number of different methods in experimental strain analysis, and in particular, to electrical resistance strain gauges and photoeleasticity.

Acknowledgements

The author would like to thank the following of his colleagues for the helpful comments and contributions they have made to him on the subject of stress analysis over many years:

Terry Johns, Harry Brown, Jim Byrne, Mike Devane, John Gibbs, Dave Hewitt, Brian Lord, Harry Newman and Phil Thompson.

In particular, the author is grateful to Professor Terry Duggan and to Graham White for the continued interest they have shown in the author's work over a period of twenty years.

Finally, he would like to thank Mrs Lesley Jenkinson for the considerable care and devotion she showed in typing the manuscript.

Notation

Unless otherwise, stated, the following symbols are used:

E	Young's modulus of elasticity
F	shearing force (S.F.)
G	shear or rigidity modulus
g	acceleration due to gravity
I	second moment of area
J	polar second moment of area
K	bulk modulus
l	length
M	bending moment (B.M.)
P	pressure
R, r	radius
T	torque or temperature change
t	thickness
W	concentrated load
w	load/unit length
$\hat{x}$	maximum value of x
α	coefficient of linear expansion
γ	shear strain
ε	direct or normal strain
θ	angle of twist
ν	Poisson's ratio
ρ	density
σ	direct or normal stress
τ	shear stress

KE   kinetic energy
PE   potential energy
UDL   uniformly distributed load

WD   work done
2E11 2×10^{11}
3.2E-3 3.2×10^{-3}
*   multiplier
⇒   vector defining the direction of rotation, according to the *right-hand screw rule*. The direction of rotation, according to the right-hand screw rule, can be obtained by pointing the right hand in the direction of the double-tailed arrow, and rotating it *clockwise*.

SOME SI UNITS IN STRESS ANALYSIS

s   second (time)
m   metre
kg   kilogram (mass)
N   newton (force)
Pa   pascal (pressure) = 1 N/m^2
MPa   magapascal (10^6 pascals)
bar   (pressure), where 1 bar = 10^5 N/m^2 = 14.5 lbf/in^2
kg/m^3 kilograms/cubic metre (density)
W   watt (power), where 1 watt = 1 ampere * 1 volt = 1 N m/s = 1 joule/s
h.p.   horse-power (power), where 1 h.p. = 745.7 W

AUTHOR'S NOTE ON THE SI SYSTEM

Is it not interesting to note that

$$1 \text{ W} = 1 \text{ N m/s} = 1 \text{ amp} \times 1 \text{ volt} = 1 \text{ joule/s}$$

and also that,

an *apple* weighs approximately one *newton*?

PARTS OF THE GREEK ALPHABET COMMONLY USED IN MATHEMATICS

α alpha
β beta
γ gamma
δ delta
Δ delta (capital)
ε epsilon
ζ zeta

η	eta
θ	theta
κ	kappa
λ	lambda
μ	mu
ν	nu
ξ	xi
Ξ	xi (capital)
π	pi
σ	sigma
Σ	sigma (capital)
τ	tau
ϕ	phi
χ	chi
ψ	psi
ω	omega
Ω	omega (capital)

To Ann, Nicolette, Jonathan and mum,
and to the memory of my late father.

1

Statics

1.1.1

All the structures that are analysed in the present chapter are assumed to be in *equilibrium*. This is a fundamental assumption that is made in the structural design of most structural components, some of which can then be designed quite satisfactorily from statical considerations alone. Such structures are said to be *statically determinate*.

Most modern structures, however, cannot be solved by considerations of statics alone, as there are more unknown "forces" than there are simultaneous equations, obtained from observations of statical equilibrium. Such structures are said to be *statically indeterminate*, and analysis of this class of structure will not be carried out in the present chapter.

Examples of statically determinate and statically indeterminate pin-jointed trusses are given in Figs. 1.1 and 1.2, where the applied concentrated loads and support reactions (R and H) are shown by arrows.

The structures of Figs 1.1 and 1.2 are called pin-jointed trusses, because the joints are assumed to be held together by smooth frictionless pins. The reason for this assumption is that the solution of trusses with pin-joints is considerably simpler than if the joints were assumed welded (i.e. rigid-joints). It should be noted that the external loads in Figs 1.1 and 1.2 are assumed to be applied at the joints, and providing this assumption is closely adhered to, the differences between the internal member forces determined from a pin-jointed truss calculation and those obtained from a rigid-jointed "truss" calculation will be small, as shown in Section 1.4.1.

The members of a pin-jointed truss are called *rods*, and these elements are assumed to withstand loads axially, so that they are either in tension or in compression or in a state of zero load. When a rod is subjected to tension it is

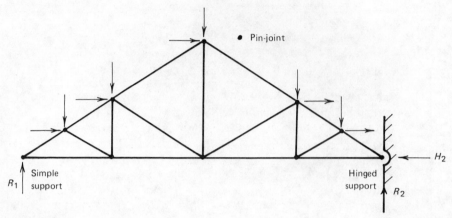

Fig. 1.1.   Statically determinate pin-jointed truss.

called a *tie* and when it is in compression, it is called a *strut*, as shown in Fig. 1.3. The internal resisting forces inside ties and struts will act in an opposite direction to the externally applied forces, as shown in Fig. 1.4. Ties, which are in *tension*, are said to have internal resisting forces of *positive magnitude*, and struts, which are in compression are said to have internal resisting forces of *negative magnitude*.

1.1.2

The structure of Fig. 1.2 is said to be statically indeterminate to the first degree, or that it has one *redundant member*; by this, it is meant that for the structure to be classified as a structure, one rod may be removed. If two rods

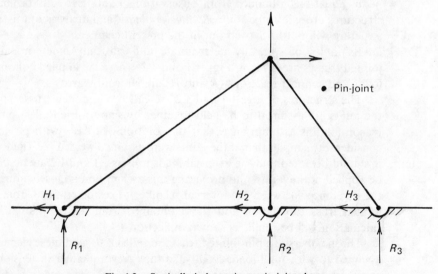

Fig. 1.2.   Statically indeterminate pin-jointed truss.

TIE (and externally applied loads)   STRUT (and externally applied loads)

Fig. 1.3.   External loads acting on ties and struts.

were removed from Fig. 1.2 or one from Fig. 1.1, the structure (or part of it) would become a *mechanism*, and collapse.

It should be noted, however, that if a rod were removed from Fig. 1.2, although the structure can still be classified as a structure and not a mechanism, it may become too weak and collapse anyhow.

TIE, with internal resisting   STRUT, with internal resisting
forces acting inwards (positive)   forces acting outwards (negative)

Fig. 1.4.   Sign conventions for ties and struts.

1.1.3   CRITERION FOR SUFFICIENCY OF BRACING

A simple formula, which will be given without proof, to test whether or not a pin-jointed truss is statically determinate, is as follows:

$$2j = r + R \tag{1.1}$$

where

j = number of pin-joints
r = number of rods
R = minimum number of reacting forces.

To test equation (1.1), consider the pin-jointed truss of Fig. 1.1.

$$j = 10, \qquad r = 17, \qquad R = 3$$

i.e.

$$2 \times 10 = 17 + 3 \text{ (OK)}$$

Now consider the truss of Fig. 1.2.

$$j = 4, \qquad r = 3, \qquad R = 6$$

or,

$$2 \times 4 = 3 + 6 \text{ (i.e. there is one redundancy)} \tag{1.2}$$

N.B. If $2j > r + R$, then the structure (or part of it) is a mechanism and will collapse under load. Similarly, from (1.2), it can be seen that if $2j < r + R$, the structure may be statically indeterminate.

1.2.1   EQUILIBRIUM CONSIDERATIONS

In *two dimensions*, the following three equilibrium considerations apply:

(a)   Vertical equilibrium must be satisfied, i.e.

Upward forces = Downward forces.

(b)   Horizontal equilibrium must be satisfied, i.e.

Forces to the left = Forces to the right.

(c)   Rotational equilibrium must be satisfied at all points, i.e.

Clockwise couples = Anti-clockwise couples.

1.3.1

To demonstrate that static analysis of pin-jointed trusses, the following example will be considered.

1.3.2   EXAMPLE 1.1   PLANE PIN-JOINTED TRUSS

Determine the internal forces in the members of the statically determinate pin-jointed truss shown in Fig. 1.5.

1.3.3

From Fig. 1.5, it can be seen that to achieve equilibrium, it will be necessary for the three reactions R_A, R_B and H_B to act. These reactions balance vertical and horizontal forces and rotational couples.

To determine the unknown reactions, it will be necessary to obtain three simultaneous equations from the three equilibrium considerations described in Section 1.2.1.

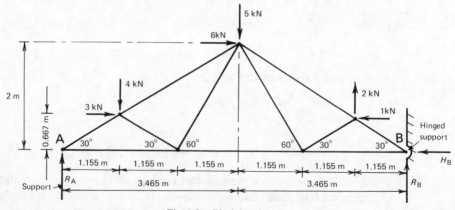

Fig. 1.5.   Pin-jointed truss.

Consider horizontal equilibrium   Forces to the left = Forces to the right, i.e.

$$H_B = 3 + 6 - 1$$

$$\underline{\underline{H_B = 8 \text{ kN}}} \tag{1.3}$$

Consider vertical equilibrium   Upward forces = Downward forces

$$R_A + R_B + 2 = 4 + 5$$

or

$$\underline{\underline{R_B = 7 - R_A}} \tag{1.4}$$

Consider rotational equilibrium   Now rotational equilibrium can be considered at any point in the plane of the truss, but to simplify arithmetic, it is better to take moments about a point through which an unknown force acts, so that this unknown force has no moment about this point. In this case, it will be convenient to *take moments* (i.e. consider rotational equilibrium), about either "A" or "B".

Taking moments about B   Clockwise couples about B = Anti-clockwise couples about B

$$R_A \times 6.93 + 3 \times 0.667 + 6 \times 2 + 2 \times 1.155$$

$$= 4 \times 5.775 + 5 \times 3.465 + 1 \times 0.667$$

i.e.

$$\underline{\underline{R_A = 3.576 \text{ kN}}} \tag{1.5}$$

Substituting (1.5) into (1.4):

$$\underline{\underline{R_B = 3.424 \text{ kN}}} \tag{1.6}$$

1.3.4

To determine the internal forces in the rods, due to these external forces, assume all *unknown member forces are in tension*. This assumption will be found convenient, because if a member is in compression, the sign of its member force will be negative, which is the correct sign for a compressive force.

The method adopted in this section for determining member forces is called the *method of joints.*

The method of joints consists of isolating each joint in turn, by making an imaginary cut through the members around that joint and then considering vertical and horizontal equilibrium between the internal member forces and the external forces acting at that joint.

As only vertical and horizontal equilibrium is considered at each joint, it is necessary to start the analysis at a joint where there are only *two unknown*

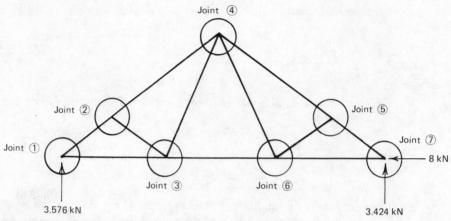

Fig. 1.6.   Joint numbers for pin-jointed truss.

member forces. In this case, consideration must first be made at either joint ①
or joint ⑦. (See Fig. 1.6.)
Joint ① (see Fig. 1.7)

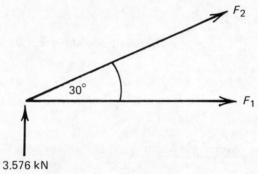

Fig. 1.7.   Joint ①.

Resolving forces vertically

Upward forces = downward forces

$$3.576 + F_2 \sin 30 = 0$$

therefore

$$F_2 = -7.152 \text{ kN (compression)} \tag{1.7}$$

Resolving forces horizontally

Forces to the left = forces to the right

$$0 = F_2 \cos 30 + F_1 \quad \text{or} \quad F_1 = -F_2 \cos 30$$

therefore

$$\underline{F_1 = 6.194 \text{ kN (tension)}} \tag{1.8}$$

Joint ② (*see Fig.* 1.8)

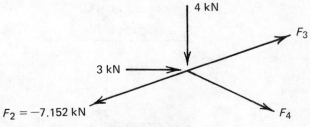

Fig. 1.8.   Joint ②.

Resolving horizontally

$$F_3 \cos 30 + F_4 \cos 30 + 3 = F_2 \cos 30$$

therefore

$$F_3 = -10.616 - F_4 \qquad (1.9)$$

Resolving vertically

$$F_3 \sin 30 = F_4 \sin 30 + F_2 \sin 30 + 4$$

or

$$F_4 = F_3 - 0.848 \qquad (1.10)$$

Substituting equation (1.9) into (1.10):

$$\underline{F_4 = -5.732 \text{ kN (comp)}} \qquad (1.11)$$

From equation (1.9):

$$\underline{F_3 = -4.884 \text{ kN (comp)}} \qquad (1.12)$$

Joint ③ (*see Fig.* 1.9)

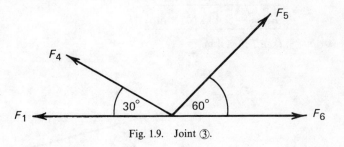

Fig. 1.9.   Joint ③.

Resolve vertically

$$F_5 \sin 60 + F_4 \sin 30 = 0$$

$$\underline{F_5 = 3.309 \text{ kN (tensile)}} \qquad (1.13)$$

Resolve horizontally

$$F_1 + F_4 \cos 30 = F_6 + F_5 \cos 60$$

$$\underline{F_6 = -0.424 \text{ kN (comp)}} \qquad (1.14)$$

Joint ④ *(see Fig. 1.10)*

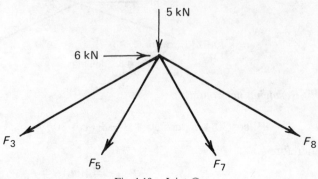

Fig. 1.10. Joint ④.

Resolve vertically

$$0 = 5 + F_3 \sin 30 + F_5 \sin 60 + F_7 \sin 60 + F_8 \sin 30$$

or

$$0.866 \, F_7 + 0.5 \, F_8 = -5.425 \qquad (1.15)$$

Resolve horizontally

$$6 + F_8 \cos 30 + F_7 \cos 60 = F_3 \cos 30 + F_5 \cos 60$$

or

$$F_8 = -0.577 \, F_7 - 9.902 \qquad (1.16)$$

Substituting equation (1.16) into (1.15):

$$\underline{F_7 = -0.82 \text{ kN (comp)}} \qquad (1.17)$$

$$\underline{F_8 = -9.43 \text{ kN (comp)}} \qquad (1.18)$$

Joint ⑤ *(see Fig. 1.11)*

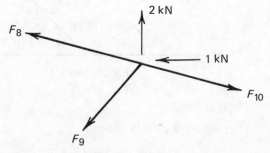

Fig. 1.11. Joint ⑤.

Resolve vertically

$$2 + F_8 \sin 30 = F_{10} \sin 30 + F_9 \sin 30$$

or

$$F_9 + F_{10} = -5.43 \tag{1.19}$$

Resolve horizontally

$$F_8 \cos 30 + F_9 \cos 30 + 1 = F_{10} \cos 30$$
$$-9.43 + F_9 + 1.155 = F_{10}$$

or

$$F_{10} - F_9 = -8.28 \tag{1.20}$$

Add (1.19) and (1.20):

$$2F_{10} = -13.7$$

Therefore

$$\underline{F_{10} = -6.85 \text{ kN (comp)}} \tag{1.21}$$

Substituting (1.21) into (1.20):

$$\underline{F_9 = 1.43 \text{ kN (tensile)}} \tag{1.22}$$

Joint ⑦ (see Fig. 1.12)

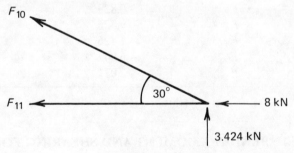

Fig. 1.12.   Joint ⑦.

Only one unknown force is required, namely F_{11}; therefore it will be easier to consider joint ⑦, rather than joint ⑥.

Resolving horizontally

$$F_{11} + 8 + F_{10} \cos 30 = 0$$

therefore

$$\underline{F_{11} = -2.07 \text{ (comp)}} \tag{1.23}$$

1.4.1   EXAMPLE 1.2   PIN-JOINTED VERSUS RIGID-JOINTED FRAMES

Using the computer programs "TRUSS" and "PLANEFRAME" of reference [1], determine the member forces in the plane pin-jointed truss of Fig. 1.5.

The program "TRUSS" assumes that the frame has pin-joints, and the program "PLANEFRAME" assumes that the frame has rigid (or welded) joints; the results are given in Table 1.1, where for "PLANEFRAME", the cross-sectional areas of the members were assumed to be 1000 times greater than their second moments of area.

As it can be seen from Table 1.1, the assumption that the joints are pinned gives similar results to the much more difficult problem of assuming that the joints are rigid (or welded), providing, of course, that the externally applied loads are at the joints.

Table 1.1.   Member forces in framework.

Force	Pin-jointed truss (kN)	Rigid-jointed frame (kN)
F_1	6.194	6.043
F_2	− 7.152	− 7.021
F_3	− 4.884	− 4.843
F_4	− 5.730	− 5.601
F_5	3.309	3.258
F_6	− 0.424	− 0.417
F_7	− 0.823	− 0.868
F_8	− 9.426	− 9.384
F_9	1.426	1.421
F_{10}	− 6.847	− 6.829
F_{11}	− 2.071	− 2.086

1.5.1   BENDING MOMENT AND SHEARING FORCE

This is of much importance in the analysis of beams and rigid-jointed frameworks, but the latter structures will not be considered in the present text, beacuse they are much more suitable for computer analysis [1].

Once again, as in section 1.2.1, all the beams will be assumed to be statically determinate and in equilibrium.

1.5.2   Definition of Bending Moment (*M*)

A *bending moment*, *M*, acting at any particular section on a beam, in equilibrium, can be defined as the resultant of all the couples acting on one side of the beam at that particular section. The resultant of the couples acting

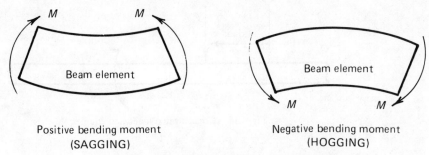

Positive bending moment
(SAGGING)

Negative bending moment
(HOGGING)

Fig. 1.13.   Sagging and hogging bending action.

on either side of the appropriate section can be considered, as the beam is in equilibrium, so that the beam will either be in a *sagging* condition or in a *hogging* one, at that section, as shown in Fig. 1.13.

The sign convention for bending moments will be that a *sagging moment* is assumed to be *positive* and a hogging one is assumed to be negative. The units for bending moment are Nm, kN m, etc.

1.5.3   Definition of Shearing Force (F)

A shearing force, F, acting at any particular section on a *horizontal beam*, in equilibrium, can be defined as the resultant of the *vertical forces* acting on one side of the beam at that particular section. The resultant of the vertical forces acting on either side of the appropriate section can be considered, as the beam is in equilibrium, as shown in Fig. 1.14, which also shows the sign conventions for positive and negative shearing forces. The units for shearing force are N, kN, MN, etc.

1.5.4

Prior to analysing beams it will be necessary to describe the symbols used for loads and supports and the various types of beam.

1.6.1   LOADS

These can take various forms, including concentrated loads and couples, which are shown as arrows in Fig. 1.15, and distributed loads, as shown in Figs 1.16 to 1.18.

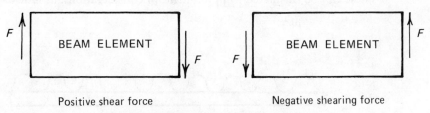

Positive shear force

Negative shearing force

Fig. 1.14.   Positive and negative shearing forces.

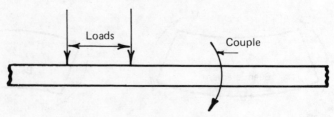

Fig. 1.15.   Concentrated loads and couples.

1.6.2

Concentrated loads are assumed to act at points. This is, in general, a pessimistic assumption. Typical concentrated loads appear in the form of loads transmitted through wheels, hanging weights, etc. The units for concentrated loads are N, kN, MN, etc.

1.6.3

Uniformly distributed loads are assumed to be distributed uniformly over the length or part of the length of a beam. Typical uniformly distributed loads are due to wind load, snow load, self-weight, etc. The units for uniformly distributed loads are N/m, kN/m, etc.

1.6.4

Hydrostatic (or trapezoidal) loads are assumed to increase or decrease linearly with length, as shown in Fig. 1.17. They usually appear in containers carrying liquids or in marine structures, etc., which are attempting to contain the water, etc. The units of hydrostatic load are N/m, kN/m, etc.

1.6.5

Varying distributed loads do not have the simpler shapes of uniformly distributed and hydrostatic loads described earlier. Typical cases of varying distributed loads are those due to the self-weight of a ship, together with the buoyant forces acting on its hull.

1.6.6   Wind Loads

These can be calculated from momentum considerations, as follows:

$$p = \rho v^2$$

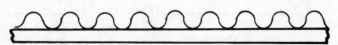

Fig. 1.16.   Uniformly distributed load (acting downwards).

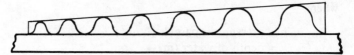

Fig. 1.17.   Hydrostatic load (acting downwards).

where,

ρ = density (kg/m^3)
p = pressure in pascals (N/m^2)
v = wind speed in m/s

This simple expression assumes that the wind acts on a flat surface and that the wind is turned through 90° from its original direction, so that the calculation for p is, in general, overestimated, and on the so-called "safe" side.

For *air* at standard temperature and pressure (*s.t.p.*),

$$\rho \fallingdotseq 1.293 \text{ kg/m}^3$$

so that, when,

$$v = 44.7 \text{ m/s (100 miles/h)}$$

$$p = 2584 \text{ Pa (0.375 p.s.i.)}$$

For an *atomic blast*,

$$p = 0.103 \text{ MPa (15 p.s.i.)}$$

so that,

$$v = 283 \text{ m/s (633 miles/h)}$$

N.B.   For a standard living room of dimensions, say, 5 m × 5 m × 2.5 m, the air contained in the room will weight about 80.8 kg (178.1 lbf), and for a large hall of dimensions, say, 30 m × 30 m × 15 m, the air contained in the hall will weight about 17 456 kg (17.1 tons)!

1.6.7   Water Pressure

In general, the *pressure due to water increases linearly with the depth of the water*, according to the expression

$$p = \rho g h$$

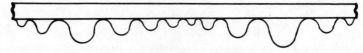

Fig 1.18.   A varying distributed load (acting upwards).

where,

p = pressure in pascals (N/m^2)

h = depth of water (m)

g = acceleration due to gravity (m/s^2)

ρ = density of water (kg/m^3), which for pure water, at normal temperature and pressure = 1000 kg/m^3, and for sea water = 1020 kg/m^3

At a *depth of* 1000 *m* (in sea water)

$p \fallingdotseq$ 10 MPa (1451 p.s.i.)

so that, for a submarine of average diameter 10 m and of length 100 m, the total load, due to water pressure, will be about 37 700 MN (3.78E6 tons)!

N.B.   1 MN $\fallingdotseq$ 100 tons

1.7.1   TYPES OF BEAM

The simplest types of beam are those shown in Figs 1.19 and 1.20.

1.7.2

The beam of Fig. 1.19, which is statically determinate, is supported on two knife edges or *simple supports*. In practice, however, a true knife-edge support is not possible, as such edges will not have zero area.

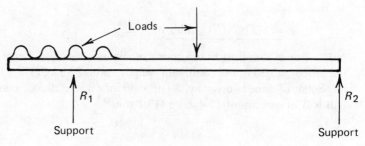

Fig. 1.19.   Statically determinate beam on simple supports.

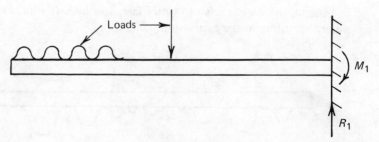

Fig. 1.20.   Cantilever.

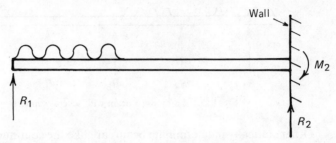

Fig. 1.21.   Propped cantilever.

1.7.3

The beam of Fig. 1.20, which is also statically determinate, is rigidly fixed (encastré) at its right end and it is called a *cantilever*.

1.7.4

It can be seen from Figs 1.19 and 1.20, that in both cases, there are two reacting "forces", which in the case of Fig. 1.19 are R_1 and R_2 and in the case of Fig. 1.20 are the vertical reaction R_1 and the restraining couple M_1. These beams are said to be statically determinate, because the two unknown reactions can be found from two simultaneous equations, which can be obtained by resolving forces vertically and taking moments.

1.7.5

Statically indeterminate beams cannot be analysed through simple statical considerations alone, because for such beams, the number of unknown "reactions" are more than the equations that can be derived from statical observations. For example, the propped cantilever of Fig 1.21 has one redundant reacting "force" (i.e. either R_1 or M_2), and the beam of Fig. 1.22, which is encastré at both ends, has two redundant reacting "forces" (i.e. either M_1 and M_2 or R_1 and M_1 or R_2 and M_2). These "forces" are said to be redundant, because if they did not exist, the structure can still be classified as a structure, as distinct from a mechanism. (N.B. by an "encastré end", it is meant that all movement, including rotation, is prevented at that end.)

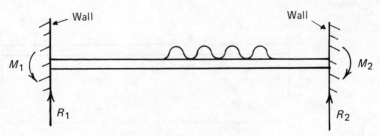

Fig. 1.22.   Beam with encastré ends.

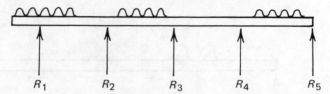

Fig. 1.23.   Continuous beam with several supports.

Other statically indeterminate beams are like the continuous beam of Fig. 1.23, which has three redundant reacting forces (i.e. it is statically indeterminate to the third degree).

To demonstrate bending moment and shearing force action, Example 1.3 will be considered, but on this occasion, the shearing forces and bending moments acting on both sides of any particular section will be calculated. The reason for doing this is to demonstrate the nature of bending moment and shearing force.

1.8.1   EXAMPLE 1.3   SIMPLY-SUPPORTED BEAM WITH CONCENTRATED LOAD

Determine expressions for the bending moment and shearing force distributions along the length of the beam of Fig. 1.24. Hence, or otherwise, plot these bending moment and shearing force distributions.

First, it will be necessary to calculate the reactions R_A and R_B, and this can be done by taking moments about a suitable point and resolving forces vertically.

1.8.2   Take Moments about B

It is convenient to take moments about either A or B, as this will eliminate the moment due to either R_A or R_B, respectively.

$$\text{Clockwise moments} = \text{anti-clockwise moments}$$

$$R_A \times 3\text{m} = 4\,\text{kN} \times 2\text{m}$$

therefore

$$\underline{R_A = 2.667\,\text{kN}}$$

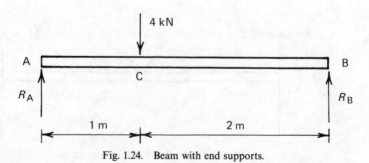

Fig. 1.24.   Beam with end supports.

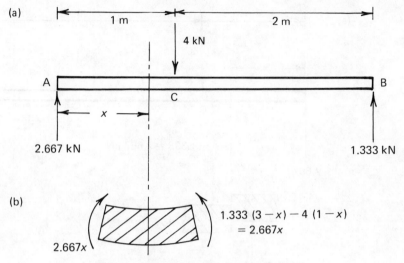

Fig. 1.25.   Bending moment between A and C.

1.8.3   Resolve Forces Vertically

Upward forces = downward forces

$$R_A + R_B = 4 \text{ kN}$$

therefore

$$\underline{R_B = 1.333 \text{ kN}}$$

1.8.4   To Determine the Bending Moment Between A and C

Consider any distance x between A and C, as shown in Fig. 1.25. From Fig. 1.25, it can be seen that at any distance x, between A and C, the reaction R_A causes a bending moment $= R_A x$, which is sagging (i.e. positive). It can also be seen from Fig. 1.25, that the forces to the right of the beam cause an equal and opposite moment, so that the beam bends at this point in the manner shown in Fig. 1.25(b).

Therefore

$$\text{Bending moment} = M = 2.667x \qquad (1.24)$$

Equation (1.24) can be seen to be a straight line, which increases linearly from zero at A to a maximum value of 2.667 kN m at C.

1.8.5   To Determine the Shearing Force Between A and C

Consider any distance x between A and C, as shown in Fig. 1.26. From Fig. 1.26(b), it can be seen that the vertical forces on the left of the beam tend to cause the left part of the beam at x to "slide" upwards, whilst the vertical

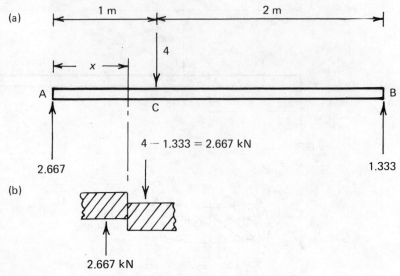

Fig. 1.26.　Shearing force between A and C.

forces on the right of the beam tend to cause the right part of the beam at x to "slide" downwards, i.e.

$$\text{shearing force at } x = F = 2.667 \text{ kN} \tag{1.25}$$

Equation (1.25) shows the shearing force to be constant between A and C, and is said to be positive, because the right side tends to move downwards, as shown in Fig. 1.26(b).

1.8.6　To Determine the Bending Moment Between C and B

Consider any distance x between C and B, as shown in Fig. 1.27. From Fig. 1.27, it can be seen that at any distance x, the bending moment =

$$M = 4 - 1.333x \text{ (sagging)} \tag{1.26}$$

Equation (1.26) shows the bending moment distribution between C and B to be decreasing linearly from 2.667 kN m at C to zero at B.

1.8.7　To Determine the Shearing Force Between C and B

Consider any distance x between C and B, as shown in Fig. 1.28. From Fig. 1.28, it can be seen that the shearing force =

$$F = -1.333 \text{ kN} \tag{1.27}$$

The shearing force F is constant between C and B, and it is negative because the right side of any section tends to "slide" upwards and the left side to "slide" downwards, as shown in Fig. 1.28(b).

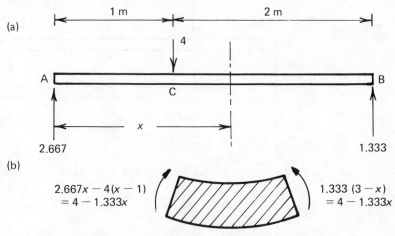

Fig. 1.27.   Bending moment between C and B.

1.8.8   Bending Moment and Shearing Force Diagrams

To obtain the bending moment diagram, it is necessary to plot equations (1.24) and (1.26), in the manner shown in Fig. 1.29(b), and to obtain the shearing force diagram, it is necesaary to plot equations (1.25) and (1.27) in the manner shown in Fig. 1.29(c).

1.8.9

From the calculations carried out in Sections 1.8.4 to 1.8.7, it can be seen that to determine the bending moment or shearing force at any particular section,

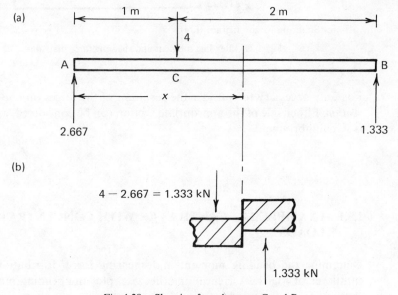

Fig. 1.28.   Shearing force between C and B.

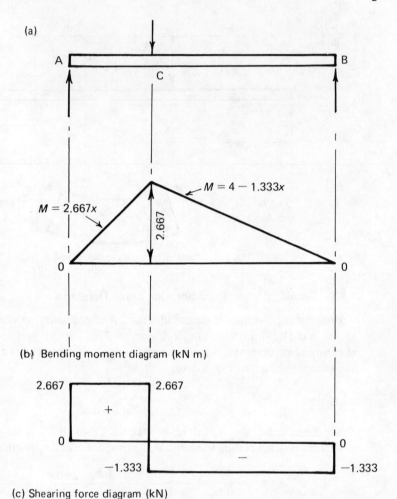

(a)

M = 2.667x

$M = 4 - 1.333x$

2.667

0   0

(b) Bending moment diagram (kN m)

(c) Shearing force diagram (kN)

Fig. 1.29.   Bending moment and shearing force diagrams.

it is only necessary to consider the resultant of the forces on *one side of the section*. Either side of the appropriate section can be considered, as the beam is in equilibrium.

1.9.1   EXAMPLE 1.4   CANTILEVER WITH CONCENTRATED LOADS

Determine the bending moment and shearing force distributions for the cantilever of Fig. 1.30. Hence, or otherwise, plot the bending moment and shearing force diagrams.

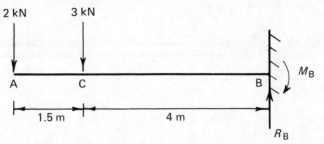

Fig. 1.30.   Cantilever.

1.9.2   To determine R_B and R_B

Resolving forces vertically

$$R_B = 2 + 3 = 5 \text{ kN}$$

Taking moments about **B**

$$M_B = 2 \times 5.5 + 3 \times 4$$

$$M_B = 23 \text{ kN m}$$

1.9.3   To Determine Bending Moment Distributions

1.9.4   Consider Span AC

At any given distance x from A, the forces to the left cause a hogging bending moment of $2x$, as shown in Fig. 1.31, i.e.

$$M = -2x \tag{1.28}$$

Equation (1.28) can be seen to increase linearly in magnitude, from zero at the free end of 3 kN m at C.

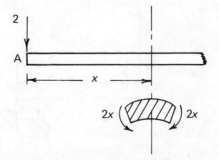

Fig. 1.31.   Bending moment between A and C.

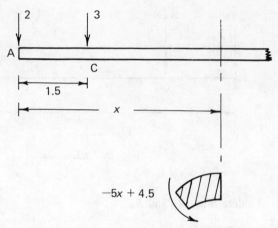

Fig. 1.32. Bending moment between C and B.

1.9.5 Consider Span CB

At any distance x from A, the forces to the left cause a hogging bending moment of $2x + (3(x - 1.5))$, as shown in Fig. 1.32, i.e.

$$M = -2x - 3(x - 1.5)$$

$$M = -5x + 4.5 \tag{1.29}$$

Equation (1.29) can be seen to increase linearly in magnitude, from 3 kN m at C to 23 kN m at B.

1.9.6 To Determine the Shearing Force Distributions

1.9.7 Consider Span AC

At any distance x, the resultant of the vertical forces to the left of this section causes a shearing force =

$$F = -2 \, \text{kN} \tag{1.30}$$

as shown in Fig. 1.33.

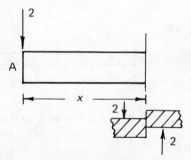

Fig. 1.33. Shearing force between A and C.

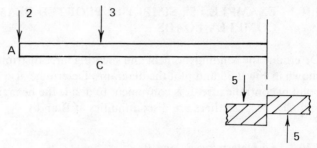

Fig. 1.34.   Shearing force between C and B.

1.9.8 Consider span CB

At any given distance x, the resultant of the vertical forces to the left of this section causes a shearing force =

$$F = -2 - 3 = -5\,\text{kN} \tag{1.31}$$

as shown in Fig. 1.34.

1.9.9 Bending Moment and Shearing Force Diagrams

Plots of the bending moment and shearing force distributions, along the length of the cantilever, are shown in Fig. 1.35.

(a)

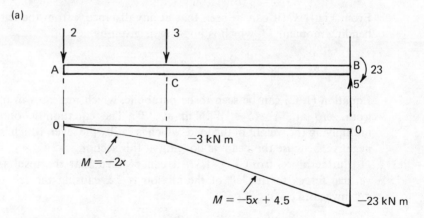

(b) Bending moment diagram (kN m)

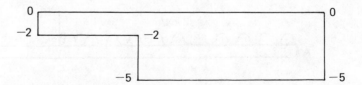

(c) Shearing force diagram (kN)

Fig. 1.35.   Bending moment and shearing force diagrams.

1.10.1   EXAMPLE 1.5   SIMPLY-SUPPORTED BEAM WITH COMPLEX LOADS

Determine the bending moment and shearing force distributions for the beam shown in Fig. 1.36 and plot the diagrams. Determine also the position of the point of contraflexure. It is convenient to divide the beam into spans AB, BC and CD, because there are discontinuities at B and C.

1.10.2   To Determine R_B and R_D

Taking moments about D

$$R_B \times 4 = 10 \times 1 + 5 \times 2 \times 3.5$$

$$\underline{R_B = 11.25 \text{ kN}}$$

Resolving forces vertically

$$R_B + R_D = 10 + 5 \times 2$$

$$\underline{R_D = 8.75 \text{ kN}}$$

1.10.3   Consider Span AB

From Fig. 1.37, it can be seen that at any distance x from the left end, the bending moment M, which is hogging, is given by,

$$M = -2 * x * \frac{x}{2} = -x^2 \qquad (1.32)$$

Equation (1.32) can be seen to be parabolic, which increases in magnitude from zero at "A" to -4 kN m at "B". The equation is obtained by multiplying the weight of the load, which is $2x$, by the lever, which is $x/2$. It is negative, beacuse the beam is hogging at this section.

Furthermore, from Fig. 1.37, it can be seen that the resultant of the vertical forces to the left of the section is $2x$, causing the left to "slide" downwards, i.e.

$$\text{shearing force} = F = -2x \qquad (1.33)$$

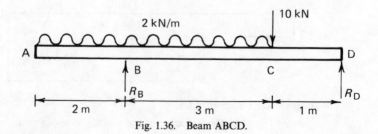

Fig. 1.36.   Beam ABCD.

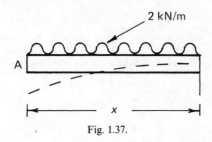

Fig. 1.37.

Equation (1.33) is linear, increasing in magnitude from zero at "A" to 4 kN at "B".

1.10.4   Consider Span BC

At any distance x on Fig. 1.38,

$$M = -2 * x * \frac{x}{2} + 11.25 (x - 2)$$

or

$$M = -x^2 + 11.25x - 22.5 \text{ (parabolic)} \qquad (1.34)$$

@ $\quad$ $x = 2, M_B = -4 \text{ kN m}$

@ $\quad$ $x = 5, M_C = 8.75 \text{ kN m}$

At any distance x,

$$F = 11.25 - 2 * x$$

$$F = 11.25 - 2x \text{ (linear)} \qquad (1.35)$$

@ $\quad$ $x = 2, F_B = 7.25 \text{ kN}$

@ $\quad$ $x = 5, F_C = 1.25 \text{ kN}$

1.10.5   Consider Span CD

At any distance x on Fig. 1.39,

$$M = 11.25(x - 2) - 2 * 5 * (x - 2.5) - 10(x - 5)$$

$$M = 52.5 - 8.75x \text{ (linear)} \qquad (1.36)$$

@ $\quad$ $x = 5, M_C = 8.75 \text{ kN m}$

@ $\quad$ $x = 6, M_D = 0 \text{ (as required)}$

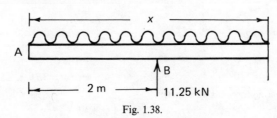

Fig. 1.38.

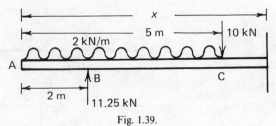

Fig. 1.39.

At any distance x in Fig. 1.39,

$$F = 11.25 - 2 \times 5 - 10$$

$$\underline{F = -8.75 \text{ kN}} \text{ (constant)} \tag{1.37}$$

1.10.6 Bending Moment and Shearing Force Diagrams

From equations (1.32) to (1.37), the bending moment and shearing force diagrams can be plotted, as shown in Fig. 1.40.

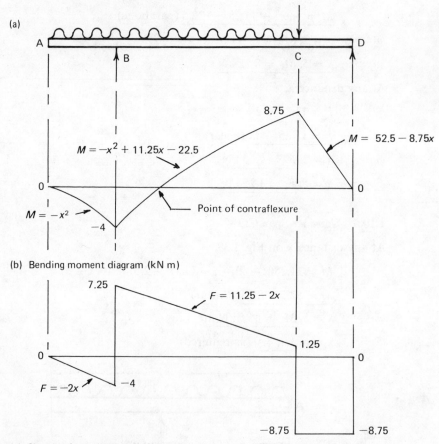

Fig. 1.40.

1.10.7   Point of Contraflexure

The *point of contraflexure* is the point on a beam where the bending moment changes sign from a positive value to a negative one, or vice versa, as shown in Fig. 1.40(b).

For this case, M must be zero between B and C, i.e. equation (1.34) must be used, so that,

$$-x^2 + 11.25x - 22.5 = 0$$

or

$$x = \frac{-11.25 + \sqrt{11.25^2 - 4 \times 22.5}}{-2}$$

$$\underline{x = 2.6 \text{ m}}$$

i.e. the point of contraflexure is 2.6 m from A or 0.6 m to the right of B.

1.11.1   EXAMPLE 1.6   BEAM WITH COUPLES

Determine the bending moment and shearing force diagrams for the simply-supported beam of Fig. 1.41, which is acted upon by a clockwise couple of 3 kN m at B and an anti-clockwise couple of 5 kN m at C.

1.11.2   To Determine R_A and R_D

Take moments about D

$$R_A \times 4 + 3 = 5$$

$$\underline{R_A = 0.5 \text{ km}}$$

Resolve vertically

$$R_A + R_D = 0$$

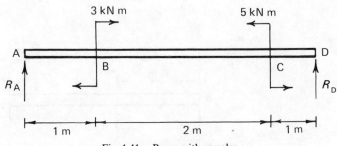

Fig. 1.41.   Beam with couples.

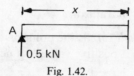

Fig. 1.42.

therefore

$$R_D = -0.5 \text{ kN}$$

i.e. R_D acts vertically downwards.

1.11.3 To Determine the Bending Moment and Shearing Force Distributions

1.11.4 Consider Span AB

At any distance x on Fig. 1.42,

$$\underline{M = 0.5x} \text{ (sagging)} \tag{1.38}$$

@   A, $\underline{M_A = 0}$

@   B, $\underline{M_B = 0.5 \text{ kN m}}$

Similarly, considering vertical forces only,

$$\underline{F = 0.5 \text{ kN}} \text{ (constant)} \tag{1.39}$$

1.11.5 Consider Span BC

At any distance x on Fig. 1.43,

$$\underline{M = 0.5x + 3} \text{ (sagging)} \tag{1.40}$$

@   B, $\underline{M_B = 3.5 \text{ kN m}}$

@   C, $\underline{M_C = 4.5 \text{ kN m}}$

Similarly, considering vertical forces only,

$$\underline{F = 0.5 \text{ kN}} \text{ (constant)} \tag{1.41}$$

1.11.6 Consider Span CD

At any distance x on Fig. 1.44,

$$M = 0.5x + 3 - 5$$

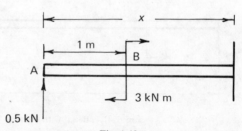

Fig. 1.43.

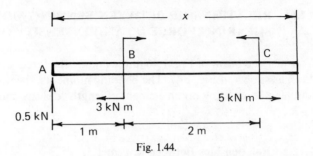

Fig. 1.44.

or

$$M = 0.5x - 2 \text{ (hogging)} \tag{1.42}$$

@ $\text{C}, M_C = -0.5 \text{ kN m}$

@ $\text{D}, M_D = 0$

Similarly, considering vertical forces only,

$$F = 0.5 \text{ kN (constant)} \tag{1.43}$$

1.11.7   Bending Moment and Shearing Force Diagram

The bending moment and shearing force distributions can be obtained by plotting equations (1.38) to (1.43), as shown in Fig. 1.45.

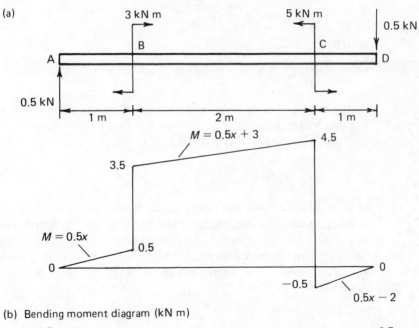

(b) Bending moment diagram (kN m)

(c) Shearing force diagram (kN)

Fig. 1.45.   Bending moment and shearing force diagrams.

1.12.1 RELATIONSHIP BETWEEN BENDING MOMENT (M), SHEARING FORCE (F) AND INTENSITY OF LOAD (w)

Consider the beam of Fig. 1.46 and examine the forces acting on an element "dx", as shown in Fig. 1.47. The bending moments, shearing forces and load/unit length, w, acting on an element of length "dx" are shown positive in Fig. 1.47.

1.12.2 Relationships Between w, F and M

Take moments about the right side of the element of Fig. 1.47, as this will eliminate "dF".

$$M + F dx = M + dM$$

or

$$\frac{dM}{dx} = F \tag{1.44}$$

i.e. the derivative of the bending moment with respect to x is equal to the shearing force at x.

Resolving forces vertically

$$F + w dx = F + dF$$

therefore

$$\frac{dF}{dx} = w \tag{1.45}$$

i.e. the derivative of the shearing force with respect to x is equal to w, the load per unit length, at x.

From equations (1.44) and (1.45)

$$\frac{d^2 M}{dx^2} = w \tag{1.46}$$

From equations (1.44) and (1.46), it can be seen that if w, the load per unit length, is known, the shearing force and bending moment distributions can be determined through repeated integration and the appropriate substitution of boundary conditions. Concentrated loads, however, present a problem, but, in general, these can be approximated by either rectangles or trapeziums or triangles.

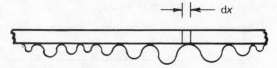

Fig. 1.46.   Beam with (positive) distributed load.

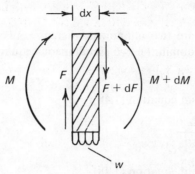

Fig. 1.47.   Beam element.

1.13.1   EXAMPLE 1.7   BEAM WITH HYDROSTATIC LOAD

Determine the shearing force and bending moment distributions for the hydrostatically loaded beam of Fig. 1.48, which is simply-supported at its ends. Find, also, the position and value of the maximum value of bending moment, and plot the bending moment and shearing force diagrams.

At any distance x, the load per unit length $=$

$$w = -1 - 0.333x$$

Now,

$$\frac{dF}{dx} = w$$

therefore

$$F = -x - 0.16667x^2 + A \tag{1.47}$$

Now,

$$\frac{dM}{dx} = F$$

therefore

$$M = -0.5x^2 - 0.05556x^3 + Ax + B \tag{1.48}$$

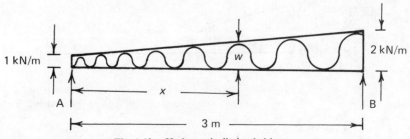

Fig. 1.48.   Hydrostatically loaded beam.

1.13.2

Now there are two unknowns; therefore, two *boundary conditions* will be required to obtain the two simultaneous equations.

@ $\qquad x = 0, M = 0$

therefore from equation (1.48):

$$\underline{B = 0}$$

@ $\qquad x_. = 3, M = 0$

therefore from equation (1.48):

$$\underline{A = 2}$$

i.e.

$$F = -x - 0.1667x^2 + 2 \qquad (1.49)$$

and

$$M = -0.5x^2 - 0.05666x^3 + 2x \qquad (1.50)$$

1.13.3   To obtain $\hat{M}$ (the maximum bending moment)

$\hat{M}$ occurs at the point where $\dfrac{\mathrm{d}M}{\mathrm{d}x} = 0$

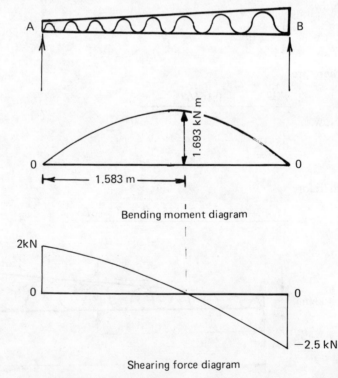

Bending moment diagram

Shearing force diagram

Fig. 1.49.   Bending moment and shearing force diagrams.

i.e.

$$-0.16667x^2 - x + 2 = 0$$

or

$$\underline{x = 1.583 \text{ m (to right of A)}} \tag{1.51}$$

Substituting equation (1.51) into (1.50):

$$\underline{\hat{M} = 1.693 \text{ kN m}}$$

1.13.4 Bending Moment and Shearing Force Diagrams

The bending moment and shearing force distributions can be obtained from equations (1.49) and (1.50), as shown in Fig. 1.49.

1.14.1 EXAMPLE 1.8 "SHIP" TYPE STRUCTURE

A barge of uniform width 10 m and length 100 m can be assumed to be of weight 40 000 kN, which is uniformly distributed over its entire length.

Assuming that the barge is horizontal and is an equilibrium, determine the bending moment and shearing force distributions, when the barge is subjected to upward buoyant forces from a wave, which is of sinusoidal shape, as shown in Fig. 1.50. The wave, whose height between peaks and trough is 3 m, may be assumed to have its peaks at the ends of the barge and its trough at the mid-length of the barge (amid-ships ⊕).

$$\rho = \text{density of water} = 1020 \text{ kg/m}^3$$

$$g = 9.81 \text{ m/s}^2$$

1.14.2 To Determine H

$$40\ 000\ 000 = 1020 \times 9.81 \times 10 \times H \times 100$$

therefore

$$\underline{H = 3.998 \text{ m or, say, 4 m}}$$

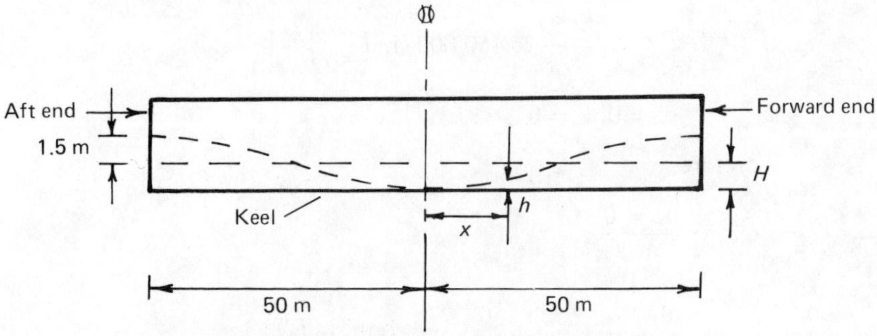

Fig. 1.50. Barge subjected to a sinusoidal wave.

@ any distance x from amid-ships, the height of the water above the keel =

$$h = 4 - 1.5 \cos\left(\frac{\pi x}{50}\right) \tag{1.52}$$

1.14.3

The *upward load* per unit length acting on the barge will be due to the buoyancy, i.e.

buoyant load/unit length at x =

$$w_b = 1020 \times 9.81 \times 10 \times \left[4 - 1.5 \cos\left(\frac{\pi x}{50}\right)\right]$$

$$w_b = 400\,000 - 150\,000 \cos\left(\frac{\pi x}{50}\right) \tag{1.53}$$

1.14.4

Now *downward load* per unit length is due to the self-weight of the base = 400 000 N/m. Therefore

$$w = w_b - 400\,000$$

$$w = -150\,000 \cos\left(\frac{\pi x}{50}\right)$$

1.14.5

Now,

$$\frac{dF}{dx} = w = -150\,000 \cos\left(\frac{\pi x}{50}\right)$$

therefore

$$F = -\frac{50}{\pi} \times 150\,000 \sin\left(\frac{\pi x}{50}\right) + A$$

@ $x = 0, F = 0$

therefore

$$A = 0$$

Now,

$$\frac{dM}{dx} = F = -\frac{50}{\pi} \times 150\,000 \sin\left(\frac{\pi x}{50}\right)$$

therefore

$$M = \left(\frac{50}{\pi}\right)^2 \times 150\,000 \cos\left(\frac{\pi x}{50}\right) + B$$

@ $x = 50$, $M = 0$

therefore

$$B = 3.8 \times 10^7$$

therefore

$$M = 3.8 \times 10^7 \left[1 + \cos\left(\frac{\pi x}{50}\right) \right] \tag{1.54}$$

and,

$$F = -2\,387\,320 \sin\left(\frac{\pi x}{50}\right) \tag{1.55}$$

$\hat{M}$ occurs at amid-ships and is equal to 76 MN m.

1.14.6 Bending Moment and Shearing Force Diagrams (Fig. 1.51)

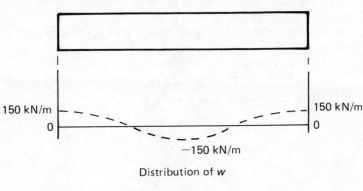

Distribution of *w*

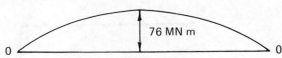

Bending moment diagram

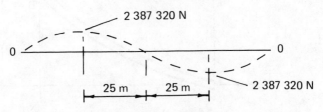

Shearing force diagram

Fig. 1.51.   Bending moment and shearing force diagrams.

1.15.1 CABLES

Cables, when acting in the mode of load carrying members, appear in a
number of different forms, varying from power lines, to cables used in
suspension bridges, and from rods used in pre-stressed concrete, to cables
used in air-supported structures. When cables are used for pre-stressed
concrete and for air-supported structures, the cables of rods are initially
placed under tension, where providing their stress values are within the
elastic limit, their bending stiffness will increase with tension. These problems,
however, which are non-linear, are beyond the scope of this book and will not
be further discussed.

Prior to analysis, it will be necessary to obtain the appropriate differential
equation that governs the deflection of cables.

1.16.1

Consider an element of cable, loaded with a distributed load, w, which is
uniform with respect to the x axis, as shown in Fig. 1.52.
Taking moments about A

$$V_2 * \mathrm{d}x = H * \mathrm{d}y + \frac{w}{2} * (\mathrm{d}x)^2 \tag{1.56}$$

Neglecting higher order terms, equation (1.56) becomes

$$V_2 = H * \frac{\mathrm{d}y}{\mathrm{d}x}\bigg|_{x=x_\mathrm{B}} \tag{1.57}$$

Similarly, by *taking moments about* B,

$$H * \mathrm{d}y + V_1 * \mathrm{d}x = \frac{w}{2} * (\mathrm{d}x)^2$$

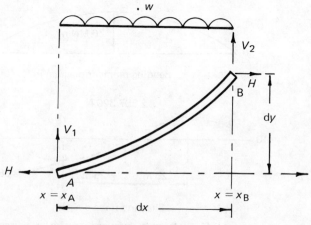

Fig. 1.52.   Cable element.

or

$$V_1 = -H * \frac{dy}{dx}\Big|_{x=x_A} \qquad (1.58)$$

Resolving vertically

$$V_1 + V_2 = w * dx$$

or

$$w = \frac{(V_1 + V_2)}{dx} \qquad (1.59)$$

Substituting equations (1.57) and (1.58) into (1.59):

$$w = \frac{H\left[\left(\dfrac{dy}{dx}\right)_{x=x_B} - \left(\dfrac{dy}{dx}\right)_{x=x_A}\right]}{dx}$$

$$= H * \frac{d^2y}{dx^2}$$

i.e.

$$\frac{d^2y}{dx^2} = \frac{w}{H} \qquad (1.60)$$

In general, where w varies with x, equation (1.60) becomes

$$\frac{d^2y}{dx^2} = \frac{w(x)}{H} \qquad (1.61)$$

where,

$w(x)$ = value of the load/unit length at any distance x

1.16.2   CABLE UNDER SELF-WEIGHT

Consider an infinitesimally small length of cable, under its own weight, as shown in Fig. 1.53. As the element is infinitesimal,

$$(ds)^2 = (dx)^2 + (dy)^2$$

or,

$$ds = dx\left[1 + \left(\frac{dy}{dx}\right)^2\right]^{1/2} \qquad (1.62)$$

Let,

w_s = weight/unit length of cable in the s direction

$w(x)$ = weight/unit length of cable in the x direction, at any distance x

$$= w_s\left[1 + \left(\frac{dy}{dx}\right)^2\right]^{1/2} \qquad (1.63)$$

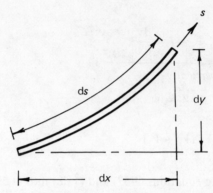

Fig. 1.53. Infinitesimal length of cable, under self-weight.

Substituting equation (1.63) into equation (1.61):

$$\frac{d^2y}{dx^2} = \frac{w_s}{H}\left[1 + \left(\frac{dy}{dx}\right)^2\right]^{1/2} \tag{1.64}$$

Solution of equation (1.64) can be achieved by letting

$$\frac{dy}{dx} = Y \tag{1.65}$$

and

$$\frac{d^2y}{dx^2} = \frac{dY}{dx} \tag{1.66}$$

Substituting equation (1.65) into (1.64),

$$\frac{dY}{dx} = \frac{w_s}{H}(1 + Y^2)^{1/2}$$

or

$$\frac{dY}{(1 + Y^2)^{1/2}} = \frac{w_s}{H} * dx$$

which, by inspection, yields the following solution:

$$\sinh^{-1}(Y) = \frac{w_s}{H} * x + C_1$$

or

$$Y = \sinh\left(\frac{w_s}{H} \cdot x + C_1\right) \tag{1.67}$$

Substituting equation (1.65) into (1.67),

$$\frac{dy}{dx} = \sinh\left(\frac{w_s}{H} \cdot x + C_1\right)$$

or

$$dy = \sinh\left(\frac{w_s}{H} \cdot x + C_1\right) \cdot dx$$

therefore

$$y = \frac{H}{w_s} \cosh\left(\frac{w_s}{H} \cdot x + C_1\right) + C_2 \tag{1.68}$$

where,

C_1 and C_2 are abritrary constants, which can be obtained from boundary value considerations.

Equation (1.68) can be seen to be the equation for a *catenary*, which is how a cable deforms naturally, when it is under its own weight.

1.16.3

The solution of equation (1.68) for practical cases is very difficult, but a good approximation for the small deflection theory of cables can be obtained by assuming a parabolic variation for y, as in equation (1.69):

$$y = \frac{w}{2H} \cdot x^2 + C_1 x + C_2 \tag{1.69}$$

where,

y = the deflection of the cable at any distance x. (See Fig. 1.54.)

C_1 and C_2 are arbitrary constants, which can be determined from boundary value considerations.

w = load/unit length in the x direction.

1.16.4

To illustrate the solution of equation (1.69), consider the cable of Fig. 1.54, which is supported at the same level at its ends.

Let,

T = tension in the cable at any distance x

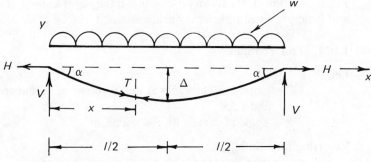

Fig. 1.54.   Cable with a uniformly distributed load.

1.16.5   To Determine C_1 and C_2

At

$$x = 0, y = 0$$

therefore

$$\underline{C_2 = 0} \tag{1.70}$$

Now,

$$\frac{dy}{dx} = \frac{w}{H} \cdot x + C_1$$

At $x = l/2$,

$$\frac{dy}{dx} = 0$$

therefore

$$0 = \frac{wl}{2H} + C_1$$

therefore

$$C_1 = -\frac{wl}{2H} \tag{1.71}$$

Substituting equations (1.70) and (1.71) into (1.69):

$$y = \frac{wx^2}{2H} - \frac{wlx}{2H} = \frac{w}{2H}(x^2 - lx) \tag{1.72}$$

1.16.6

To determine the sag Δ, substitute $x = l/2$ in equation (1.72), i.e.

$$\Delta = \frac{w}{2H}\left(\frac{l^2}{4} - \frac{l^2}{2}\right)$$

or

$$\text{sag} = \Delta = -\frac{wl^2}{8H} \tag{1.73}$$

From equation (1.73), it can be seen that if the sag is known, H can be found, and hence, from equilibrium considerations, V can be calculated

1.16.7   To Determine T

Let,

> T = tension in the cable at any distance x, as shown by Figs 1.54 and 1.55.
>
> θ = angle of cable with the x axis, at x

Resolving horizontally,

$$T \cos \theta = H \tag{1.74}$$

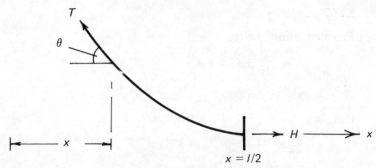

Fig. 1.55.   Element of cable.

From equation (1.74), it can be seen that the maximum tension in the cable, $\hat{T}$, will occur at its steepest gradient, so that in this case,

$$\underline{\underline{\hat{T} = \frac{H}{\cos \alpha} = H \sec \alpha}} \qquad (1.75)$$

where,

$\hat{T}$ = maximum tension in the cable, which is at the ends of the cable of Fig. 1.54.

α = slope of this cable at its ends.

1.17.1   EXAMPLE 1.9   CABLE WITH UDL

Determine the maximum tension in the cable of Fig. 1.56, where,

$$w = 120 \text{ N/m} \qquad \text{and} \qquad H = 20 \text{ kN}$$

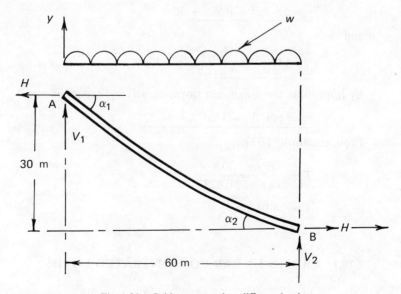

Fig. 1.56.   Cable supported at different levels.

1.17.2

From equation (1.69),

$$y = \frac{wx^2}{2H} + C_1 x + C_2 \tag{1.76}$$

@ $\qquad x = 0, y = 30$

therefore

$$\underline{C_2 = 30} \tag{1.77}$$

@ $\qquad x = 60, y = 0$

therefore

$$0 = \frac{60}{H} \cdot 60^2 + 60 \cdot C_1 + 30$$

$$C_1 = -\frac{3600}{H} - 0.5$$

but,

$$H = 20\,000$$

therefore

$$\underline{C_1 = -0.68} \tag{1.78}$$

Substituting equations (1.77) and (1.78) into (1.76),

$$y = \frac{120 \cdot x^2}{2 \times 20\,000} - 0.68x + 30$$

$$\underline{y = 3\text{E-}3x^2 - 0.68x + 30} \tag{1.79}$$

and,

$$\underline{\frac{dy}{dx} = 6\text{E-}3x - 0.68} \tag{1.80}$$

By inspection, the maximum slope, namely α_1, occurs at $x = 0$.

$$\underline{\alpha_1 = \tan^{-1}(-0.68) = -34.22°}$$

From equation (1.75):

$$T = \frac{H}{\cos \alpha_1} = \frac{20\,\text{kN}}{0.827}$$

$$\underline{\hat{T} = 24.19\,\text{kN}}$$

1.18.1 CABLES UNDER CONCENTRATED LOADS

Cables in this category are assumed to deform, as shown in Figs 1.57 to 1.59.

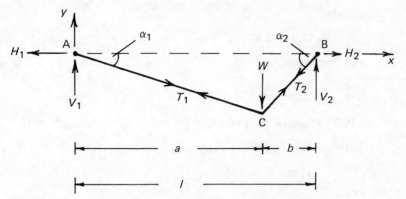

Fig. 1.57.   Cable with a single concentrated load.

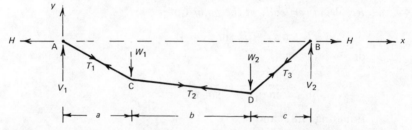

Fig. 1.58.   Cable with two concentrated loads, but with the end supports at the same level.

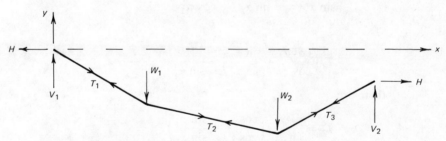

Fig. 1.59.   Cable with two concentrated loads, but with the end supports at different levels.

1.19.1   EXAMPLE 1.10   CABLE WITH A SINGLE CONCENTRATED LOAD

Determine expressions for T_1, T_2, H, V_1 and V_2, in terms of α_1 and α_2, for the cable of Fig. 1.57.

1.19.2   Resolving Vertically

$$V_1 + V_2 = W \qquad (1.81)$$

Taking moments about A

$$V_2 * l = W * a$$

therefore

$$V_2 = Wa/l$$

and from equation (1.81),

$$V_1 = Wb/l$$

1.19.3 Consider the Point C

Resolving horizontally at the point C,

$$T_1 \cos \alpha_1 = T_2 \cos \alpha_2$$

or

$$T_1 = T_2 \cos \alpha_2/\cos \alpha_1 \tag{1.82}$$

Resolving vertically at the point C,

$$T_1 \sin \alpha_1 + T_2 \sin \alpha_2 = W$$

or

$$T_1 = \frac{W}{\sin \alpha_1} - T_2 \frac{\sin \alpha_2}{\sin \alpha_1} \tag{1.83}$$

Equating (1.82) and (1.83):

$$\frac{W}{\sin \alpha_1} - T_2 \frac{\sin \alpha_2}{\sin \alpha_1} = \frac{T_2 \cos \alpha_2}{\cos \alpha_1}$$

or

$$T_2 \left(\frac{\cos \alpha_2}{\cos \alpha_1} + \frac{\sin \alpha_2}{\sin \alpha_1} \right) = \frac{W}{\sin \alpha_1}$$

therefore

$$T_2 = \frac{W}{\sin \alpha_1} \bigg/ \left(\frac{\cos \alpha_2}{\cos \alpha_1} + \frac{\sin \alpha_2}{\sin \alpha_1} \right) \tag{1.84}$$

Substituting equation (1.84) into (1.83), T_1 can be determined.

1.19.4

H_1 and H_2 can be determined by resolution, as follows.
Resolving horizontally,

$$H_1 = H_2 = T_1 \cos \alpha_1 = T_2 \cos \alpha_2 \tag{1.85}$$

1.20.1 EXAMPLE 1.11 CABLE WITH TWO CONCENTRATED LOADS

Determine the tensions T_1, T_2 and T_3 that act in the cable of Fig. 1.60. Hence, or otherwise, determine the end forces H_1, V_1, H_2 and V_2.

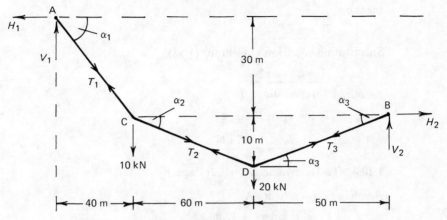

Fig. 1.60.   Cable with concentrated loads.

1.20.2   Resolving Vertically

$$V_1 + V_2 = 30 \qquad (1.86)$$

$$\alpha_1 = \tan^{-1}(0.75) = 36.87°$$

$$\alpha_2 = \tan^{-1}(10/60) = 9.46°$$

$$\alpha_3 = \tan^{-1}(10/50) = 11.31°$$

1.20.3   To Determine T_1, T_2 and T_3

Resolving horizontally at "C"

$$T_1 \cos \alpha_1 = T_2 \cos \alpha_2$$

or

$$0.8T_1 = 0.986T_2$$

therefore

$$T_1 = 1.233\, T_2 \qquad (1.87)$$

Resolving vertically at C

$$T_1 \sin \alpha_1 = 10 + T_2 \sin \alpha_2$$

$$0.6T_1 = 10 + 0.164\, T_2$$

or

$$T_1 = 16.67 + 0.273T_2 \qquad (1.88)$$

Equating (1.87) and (1.88):

$$1.233\, T_2 = 16.67 + 0.273\, T_2$$

therefore
$$T_2 = 17.37 \text{ kN} \tag{1.89}$$
Substituting equation (1.89) into (1.88),
$$T_1 = 21.42 \text{ kN} \tag{1.90}$$
Resolving horizontally at D
$$T_3 \cos \alpha_3 = T_2 \cos \alpha_2$$
$$T_3 = 17.47 \text{ kN} \tag{1.91}$$

1.20.4 To Determine H_1, V_1, H_2 and V_2

Resolving vertically at "A"
$$V_1 = T_1 \sin \alpha_1 = 12.85 \text{ kN}$$
Resolving horizontally at A
$$H_1 = T_1 \cos \alpha_1 = 17.14 \text{ kN}$$
Resolving vertically at B
$$V_2 = T_3 \sin \alpha_3 = 3.43 \text{ kN}$$
Resolving horizontally at B
$$H_2 = T_3 \cos \alpha_3 = 17.13 \text{ kN}$$
N.B. $H_1 = H_2$ (as required)

1.21.1 SUSPENSION BRIDGES

The use of cables to improve the structural efficiency of bridges is widely adopted throughout the world, especially for suspension bridges. In this case, the cables are placed under tension, between towers, so that the cables exert upward forces to support the main structure of the bridge, via vertical tie-

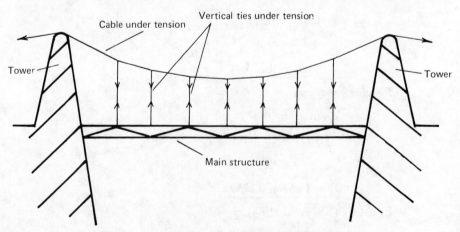

Fig. 1.61. Suspension bridge.

bars, a shown in Fig. 1.61. Most of the world's longest bridges are, in fact, suspension bridges.

EXAMPLES FOR PRACTICE 1

1. A plane pin-jointed truss is firmly pinned at its base, as shown in Fig. Q.1.1.

      Determine the forces in the members of this truss, stating whether they are in tension or compression.

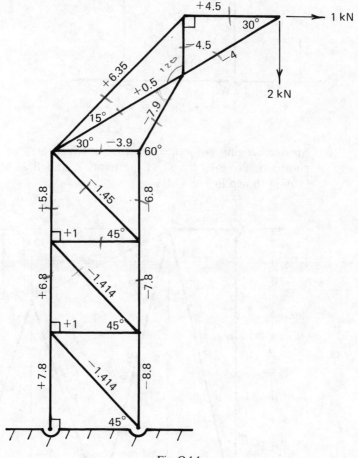

Fig. Q.1.1.

{Forces in kN; + tension; − compression}.

(Portsmouth Polytechnic, June 1980)

2. The plane pin-jointed truss of Fig. Q.1.2 is firmly pinned at A and B and subjected to two point loads at the point F.

      Determine the forces in the members, stating whether they are tensile or compressive.

(Portsmouth Polytechnic, June 1982)

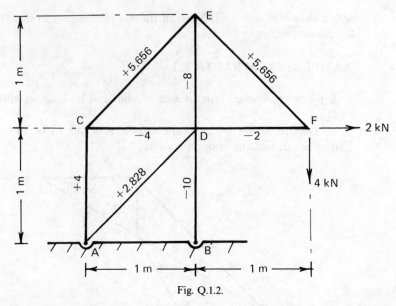

Fig. Q.1.2.

3. An overhanging pin-jointed roof truss, which may be assumed to be pinned rigidly to the wall at the joints A and B, is subjected to the loading shown in Fig. Q.1.3.

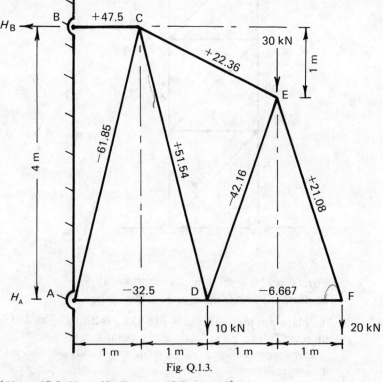

Fig. Q.1.3.

$\{H_A = 47.5, V_A = 60, H_B = -47.5, V_B = 0\}$.

Determine the forces in the members of the truss, stating whether they are tensile or compressive.

Determine, also, the reactions.

4. Determine the forces in the symmetrical pin-jointed truss of Fig. Q.1.4, stating whether they are tensile or compressive

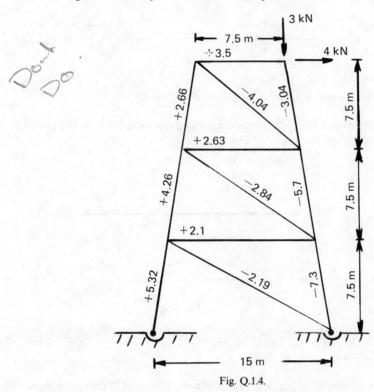

Fig. Q.1.4.

(Portsmouth Polytechnic, June 1977)

5. Determine the bending moments and shearing forces at the points A, B, C, D and E for the simply-supported beam of Fig. Q.1.5.

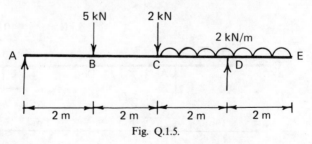

Fig. Q.1.5.

$\{M(\text{kN m}) \to 0, 8, 6, -4, 0; F(\text{kN}) \to 4, 4/-1, -1/-3, -7/4, 0; 5.73 \text{ m}$ to right of A$\}$.

Determine, also, the position of the point of contraflexure.

(Portsmouth Polytechnic, June 1977)

6. Determine the bending moments and shearing forces at the points A, B, C
   and D on the cantilever of Fig. Q.1.6.

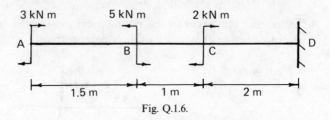

Fig. Q.1.6.

$\{M(\text{kN m}) \to 3, 3/-2, -2/0, 0; F(\text{kN}) \to 0, 0, 0, 0\}.$

7. Determine the bending moments and shearing forces at the points A, B, C
   and D on the beam of Fig. Q.1.7.

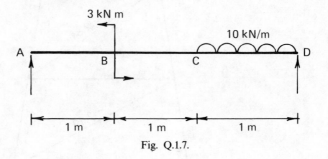

Fig. Q.1.7.

$\{M(\text{kN m}) \to 0, 2.667/-0.333, 2.333, 0; F(\text{kN}) \to 2.667, 2.667, 2.667, -7.333\}.$

8. A uniform section beam is simply-supported at A and B, as shown in Fig.
   Q.1.8. Determine the bending moments and shearing forces at the points
   C, A, D, E and B.

(Portsmouth Polytechnic, June 1982)

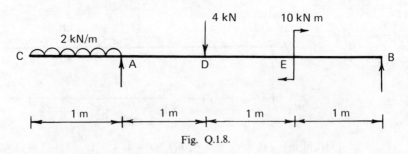

Fig. Q.1.8.

$\{M(\text{kN m}) \to 0, -1, -1.33, -5.67/4.33; 0; F(\text{kN}) \to 0, -2/0.33, 0.33/-4.33, -4.33, -4.33\}.$

9. A simply-supported beam supports a distributed load, as shown in Fig. Q.1.9.

    Obtain an expression for the value of a, so that the bending moment at the support will be of the same magnitude as that at mid-span.

(Portsmouth Polytechnic, June 1980)

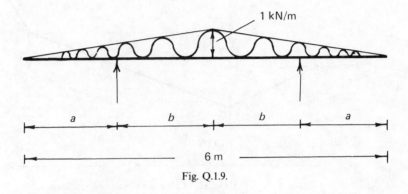

Fig. Q.1.9.

$\{a^3/18 + 1.5a = 3; a = 1.788 \text{ m}\}$.

10. Determine the maximum tensile force in the cable of Fig. Q.1.10, and the vertical reactions at its ends, given the following:

$$w = 200 \text{ N/m} \quad \text{and} \quad H = 30 \text{ kN}$$

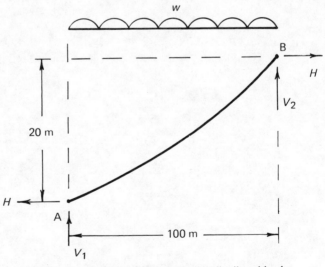

Fig. Q.1.10.   Cable with a uniformly distributed load.

$\{\hat{T}(@\ B) = 34.0 \text{ kN}; V_1 = 4 \text{ kN}; V_2 = 16 \text{ kN}\}$.

11. Determine the tensile forces in the cable of Fig. Q.1.11, together with the end reactions.

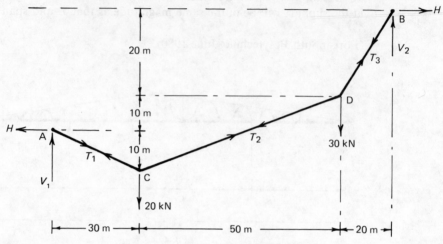

Fig. Q.1.11.   Cable with concentrated loads.

$\{T_1 = 28.76 \text{ kN}; \quad T_2 = 29.40 \text{ kN}; \quad T_3 = 38.6 \text{ kN}; \quad H = 27.28 \text{ kN}; \quad V_1 = 9.09 \text{ kN}; \quad V_2 = 27.29 \text{ kN}\}.$

2

Stress and Strain

2.1.1

The most elementary definition of stress is that it is the *load per unit area* acting on a surface, rather similar to pressure, except that it can be either tensile or compressive and it does not necessarily act normal to the surface, i.e.

$$\text{stress} = \text{load/area} \tag{2.1}$$

In its simplest form, stress acts at an angle to the surface, as shown in Fig. 2.1.

2.1.2

However, in the form shown in Fig. 2.1, it is difficult to apply stress analysis to practical problems, and because of this, the resultant stress is represented by a *normal or direct stress*, σ, together with a *shear stress*, τ, as shown in Fig. 2.2.

The stress, σ, in Fig. 2.2 is called a normal or direct stress, because it acts perpendicular to the surface under consideration, and the stress, τ, is called a

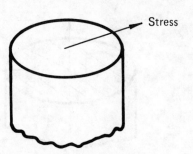

Fig. 2.1. Resultant stress.

75

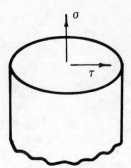

Fig. 2.2.   Normal and shear stress.

shearing stress, because it acts tangentially to the surface, causing shearing action, as shown in Fig. 2.4. Thus, if a flat surface is subjected to a force, R, acting at an angle to the surface, it is convenient to represent this resultant force by its two perpendicular components, namely P and F, where P acts normal to the surface and F acts tangentially to the surface, as shown in Fig. 2.3.

2.1.3

The effect of P will be to increase the length of the structural component and to cause a normal or direct stress, σ, where,

$$\sigma = P/A \tag{2.2}$$

where,

A = cross-sectional area

Similarly, the effect of F will be to cause the component to suffer shear deformation, as shown in Fig. 2.4, and to cause a shear stress, τ, where,

$$\tau = F/A \tag{2.3}$$

2.1.4

The *Sign Convention* for *Direct Stress* is as follows:
                Tensile stresses are positive
                Compressive stresses are negative

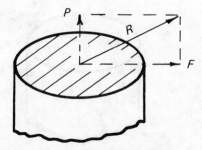

Fig. 2.3.   Components of R.

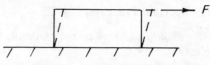

Fig. 2.4.   Shearing action of F.

2.2.1   HOOKE'S LAW

If a length of wire, made from steel or aluminium alloy, is tested in tension, the wire will be found to increase its length linearly with increase in load, for "smaller" values of load, so that,

$$\text{load} \propto \text{extension} \tag{2.4}$$

Equation (2.4) was discovered by Robert Hooke (1635–1703), and it applies to many materials up to the *limit of proportionality* of the material.

2.2.2

In structural design, Hooke's law is very important for the following reasons:

(a) In general, it is not satisfactory to allow the stress in a structural component to exceed the limit of proportionality. This is because, if the stress exceeds this value, it is likely that certain parts of the structural component will suffer permanent deformation.

(b) If the stress in a structure does not exceed the material's limit of proportionality, the structure will return to its undeformed shape on removal of the loading.

(c) For many structures it is undesirable to allow them to suffer large deformations under normal loading.

2.2.3

A typical load–extension curve for mild steel, in tension, is shown in Fig. 2.5

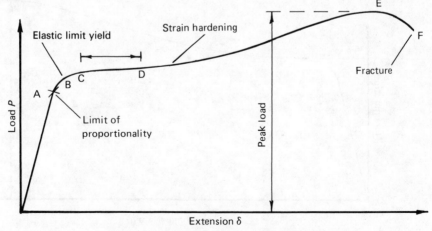

Fig. 2.5.   Load–extension curve for mild steel.

2.2.4   Some Important Points on Fig. 2.5

Limit of proportionality (A)   Up to this point, the load–extension curve is linear and elastic.

Elastic limit (B)   Up to this point, the material will recover its original shape on the removal of load. The section of the load–extension curve between the limit of proportionality and elastic limit can be described as non-linear elastic.

Yield point (C/D)   This is the point of the load–extension curve, where the body suffers permanent deformation, i.e. the material behaves plastically beyond this point, and Poisson's ratio is approximately equal to 0.5. The extension of the specimen from C to D is approximately forty times greater than the extension of the specimen up to B.

Strain hardening (D/E)   After the point D, the material strain hardens, where the slope of the load–extension curve, just above D, is about 1/50th of the slope between 0 and A.

Peak load (E)   This load is used for calculating the *ultimate tensile strength* or *tensile strength* of the material. After this point, the specimen "necks" and eventually fractures at F. (By "necking" is meant that a certain section of the specimen suffers a local decrease in its cross-sectional area.)

2.2.5   STRESS–STRAIN CURVES

Normally, Fig. 2.6 is preferred to Fig. 2.5 where,

Nominal stress, $\sigma = P/A$

Nominal strain, $\varepsilon = \delta/\text{original length}$

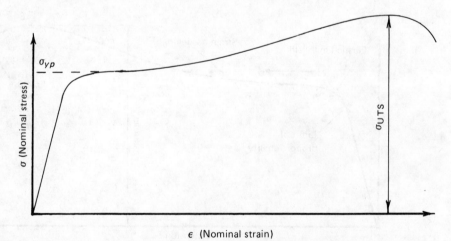

Fig. 2.6.   Nominal stress–nominal strain relationship.

where,

A = original cross-sectional area

δ = deflection due to P

On Fig. 2.6,

σ_{yp} = Yield stress, and for most structural designs, the stress–strain relationship is assumed to be linear up to this point.

σ_{UTS} = Ultimate tensile stress or nominal peak stress.

= Peak load/A

In the design of stuctures, the stress is normally not allowed to exceed the limit of proportionality, where the relationship between σ and ε can be put in the form:

$$\frac{\text{stress}(\sigma)}{\text{strain}(\varepsilon)} = E \tag{2.5}$$

where,

E = Young's (or elastic) modulus

ε = nominal strain

= extension per unit length (unitless)

2.2.6

To determine E for construction materials, such as mild steel, aluminium alloy, etc., it is normal to make a suitable specimen from the appropriate material and to load it in tension in a universal testing machine, as shown in Fig. 2.7. Prior to bending the specimen, a small extensometer is connected to the specimen, where measurements are made of the extension of the specimen over the gauge length l of the extensometer against increasing load, up to the limit of proportionality. The cross-section of the specimen is usually circular and is sensibly constant over the gauge length. In general, E is approximately the same in tension as it is in compression for most structural materials, and some typical values are given in Table 2.1.

To demonstrate some problems involving simple stress and strain, the following examples will be considered.

Table 2.1.   Young's modulus E (N/m²).

Steel	Aluminium alloy	Copper	Concrete "new"	Concrete "old"	Oak (with grain)
2.1×10^{11}	7×10^{11}	1.2×10^{11}	1.9×10^{10}	3.6×10^{10}	1.2×10^{10}

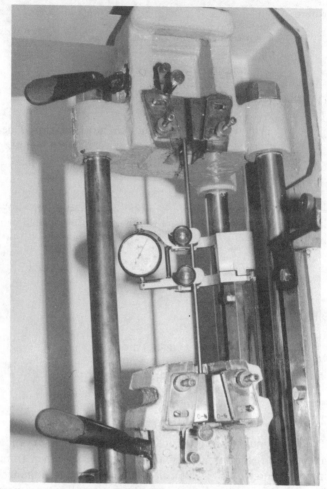

Fig. 2.7. Specimen undergoing tensile test.

2.2.7 EXAMPLE 2.1 BOLT UNDER TENSILE STRESS

Two structural components are joined together by a steel bolt with a screw thread, which has a diameter of 12.5 mm and a pitch of 1 mm, as shown in Fig. 2.8. Assuming that the bolt is initially stress free and that the structural components are inextensible, determine the stress in the bolt, if it is tightened by rotating it clockwise by $\frac{1}{8}$ of a turn. Now the *pitch* of the thread of a bolt is defined as the distance the nut attached to the bolt will move axially, if it is rotated by one complete turn. In this case, the structural components are inextensible, so that by rotating the bolt by $\frac{1}{8}$ of a turn, the bolt will increase its length by an amount δ over the length 50 mm, where,

$$\delta = \tfrac{1}{8} \times 1 \text{ mm} = 1.25 \times 10^{-4} \text{ m}$$

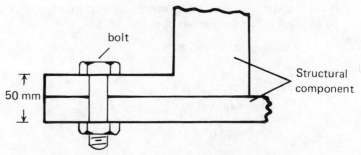

Fig. 2.8.   Bolt under tensile stress.

Now,

$$\varepsilon = \text{strain} = \frac{\delta}{l} = \frac{1.25 \times 10^{-4}}{50 \times 10^{-3}} = 2.5 \times 10^{-3}$$

$$\sigma = \text{stress} = E\varepsilon = 2 \times 10^{11} \times 2.5 \times 10^{-3}$$

$$\underline{\sigma = 500 \text{ MN/m}^2 \text{ (tensile)}}$$

2.2.8   EXAMPLE 2.2   HANGING CABLE

A copper cable hangs down a vertical mineshaft. Determine the maximum permissible length of the cable, if its maximum permissible stress, due to self-weight, must not exceed 10 MN/m². Hence, or otherwise, determine the maximum vertical deflection of this cable due to self-weight.

$$E = 1 \times 10^{11} \text{ N/m}^2 \quad \rho = 8900 \text{ kg/m}^3 \quad g = 9.81 \text{ m/s}^2$$

(a) The maximum stress in the cable will be at the top.

Let $\hat{\sigma}$ = maximum stress = 10 MN/m².
Weight of cable = $\rho A\, gl$, where

   A = cross-sectional area

   l = length of cable

   therefore $\hat{\sigma} = \dfrac{\rho A\, gl}{A} = \rho\, gl$

or,

$$10 \times 10^6 \, \frac{\text{N}}{\text{m}^2} = 8900 \times 9.81 \times l$$

i.e.

$$\underline{l = 114.5 \text{ m}}$$

(b) To determine δ, where,

   δ = maximum deflection of the cable due to self-weight

Now the stress in the cable varies *linearly* from zero at the bottom to 10 MN/m² at the top and so does the strain, i.e.

$$\text{average stress} = 5 \text{ MN/m}^2$$

and

$$\text{average strain} = \frac{5 \times 10^6}{1 \times 10^{11}} = 5 \times 10^{-5}$$

Hence,

$$\underline{\delta = 5 \times 10^{-5} \times 114.5 \text{ m} = 5.73 \text{ mm}}$$

2.2.9   EXAMPLE 2.3   CONSTANT STRESS ROD

Determine the profile of a vertical bar, which is to have a constant normal stress due to self-weight.

Consider an element of the bar at any distance x from the bottom, as shown in Fig. 2.9, where the cross-sectional area is A and the stress is σ.

Resolving vertically,

$$\sigma(A + \mathrm{d}A) = \sigma A + \rho g \, A\mathrm{d}x \tag{2.6}$$

where,

$$\rho = \text{density of material}$$

$$g = \text{acceleration due to gravity}$$

and,

$$\rho g \, A\mathrm{d}x = \text{weight of element (shaded area)}$$

Simplifying (2.6), the following is obtained:

$$\frac{\mathrm{d}A}{A} = \frac{\rho g}{\sigma} \mathrm{d}x$$

or,

$$\ln A = \frac{\rho g x}{\sigma} + \ln C \tag{2.7}$$

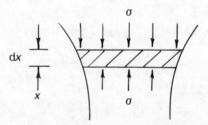

Fig. 2.9.   Constant strength rod.

where,

  ln C = an arbitrary constant

Taking the antilogarithm of equation (2.7),

  $A = Ce^{(\rho g x/\sigma)}$ (2.8)

i.e. if A is known for any x, C can be determined.

2.3.1   SHEAR STRESS AND SHEAR STRAIN

From Fig. 2.2, it can be seen that shear stress acts tangentially to the surface and that this shear stress (τ) causes the shape of a body to deform, as shown in Fig. 2.10. Although shear stress causes change of shape (or shear strain γ), it does not cause a change in volume.

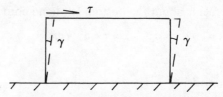

Fig. 2.10.   Shear stress τ and shear strain γ.

By experiment, it has been found that for many materials, the relationship between shear stress and shear strain is given by the equation

  $$\frac{\tau}{\gamma} = G$$ (2.9)

where,

G = modulus of rigidity or shear modulus (N/m^2, N/mm^2, MN/m^2, etc.).
γ = shear strain (unitless)

2.3.2   COMPLEMENTARY SHEAR STRESS

The effect of shearing action on an element of material, as shown in Fig. 2.11, will be to cause the system of shearing stresses of Fig. 2.12, which are assumed to be positive.

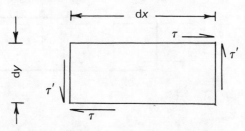

Fig. 2.11.   Shearing stresses acting on an element.

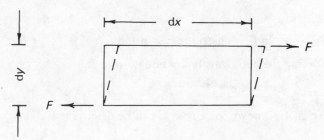

Fig. 2.12. Positive shearing stresses.

Let t = thickness of elemental lamina. By considerations of horizontal equilibrium, it is evident that τ will act on the top and bottom sides in the manner shown. The effect of these shearing stresses will be to cause a clockwise couple of $\tau * t * \mathrm{d}x * \mathrm{d}y$, and from equilibrium considerations, the system of shearing stresses τ must act in the direction shown. Hence, by taking moments about the bottom left-hand corner of the elemental lamina,

$$\tau * t * \mathrm{d}x * \mathrm{d}y = \tau' * t * \mathrm{d}y * \mathrm{d}x$$

or,

$$\tau = \tau'$$

i.e. the systems of shearing stress are complementary and equal, and positive and negative shearing stresses are shown in Fig. 2.13.

Fig. 2.13.   Positive and negative complementry shearing stresses.

2.4.1 POISSON'S RATIO (v)

If a length of wire or rubber or similar material is subjected to axial tension, as shown in Fig. 2.14, then in addition to its length increasing, its lateral dimension will decrease owing to this axial stress.

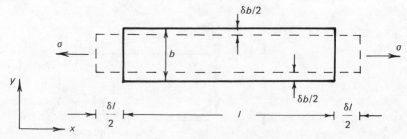

Fig. 2.14.   Axially loaded element.

The relationship between the lateral strain and the axial strain, due to the axial stress, is known as Poisson's ratio (v), where

$$v = \frac{- \text{ lateral strain}}{\text{longitudinal strain}}$$

$$\text{lateral strain} = \frac{- \delta b}{b}$$

$$\text{longitudinal strain} = \frac{\delta l}{l}$$

where,

b = breadth

δb = increment of b

l = length

δl = increment of l

From Fig. 2.14, it can be seen that although there is lateral strain, there is no lateral stress; thus, the relationship $\sigma/\varepsilon = E$ only applies for uniaxial stress in the direction of the uniaxial stress. Now,

$$\varepsilon_x = \text{longitudinal strain} = \sigma/\varepsilon$$

therefore

$$\varepsilon_y = \text{lateral strain} \qquad = - v\sigma/\varepsilon \tag{2.10}$$

Thus, for two- and three-dimensional systems of stress, equation (2.5) does not apply, and for such cases, the *Poisson effect* of equation (2.10) must also be included (see Chapter 7).

Typical values for Poisson's ratio are

0.3 for steel,   0.33 for aluminium alloy,   0.1 for concrete

It will be shown in the next section, that v cannot exceed 0.5.

2.5.1 Hydrostatic Stress

If a solid piece of material were dropped into the ocean, it would be subjected to a uniform external pressure p, caused by the weight of water above it. If an internal elemental cube from this piece of material were examined, it would be found to be subjected to the three-dimensional system of stresses, where there is no shear stress, as shown in Fig. 2.15. Such a state of stress is known as *hydrostatic stress*, and the stress everywhere is normal and equal to $- p$. If the dimensions of the cube are $x * y * z$, and the displacements due to p,

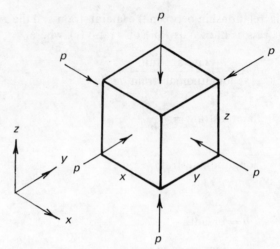

Fig. 2.15.   Elemental cube under hydrostatic stress.

corresponding to these dimensions, are δx, δy and δz, respectively, then from equation (2.10):

$$\varepsilon_x = \frac{\delta x}{x} = -\frac{p}{E}(1 - v - v)$$

$$\varepsilon_y = \frac{\delta y}{y} = -\frac{p}{E}(1 - v - v) \qquad (2.11)$$

$$\varepsilon_z = \frac{\delta z}{z} = -\frac{p}{E}(1 - v - v)$$

From equation (2.11), it can be seen that if $x = y = z$, and $v > 0.5$, then δx, δy and δz will be positive, which is impossible, i.e. *v cannot be greater than* 0.5.

In fact the stress system of Fig. 2.15 will cause the volume of the element to decrease by an amount δV, so that,

$$\text{volumetric strain} = \frac{\text{change in volume }(\delta V)}{\text{original volume }(V)}$$

$$= \frac{\delta V}{V}$$

If the material obeys Hooke's law, then,

$$\frac{\text{volumetric stress}}{\text{volumetric strain}} = K \qquad (2.12)$$

where,

$$K = \text{bulk modulus}$$

In this case, volumetric stress $= -p$.

2.5.2   To Determine Volumetric Strain

Now original volume $V = x * y * z$

New volume $V + \delta V = x(1 + \varepsilon_x) * y(1 + \varepsilon_y) * z(1 + \varepsilon_z)$
Assuming the deflections are small and neglecting higher order terms,

$$V + \delta V = xyz(1 + \varepsilon_x + \varepsilon_y + \varepsilon_z)$$

therefore

$$\delta V = xyz(\varepsilon_x + \varepsilon_y + \varepsilon_z)$$

and,

$$\text{volumetric strain} = \frac{\delta V}{V} = (\varepsilon_x + \varepsilon_y + \varepsilon_z)$$

i.e. the volumetric strain is the sum of the three co-ordinate strains ε_x, ε_y and ε_z.

2.6.1   Relationship Between the Material Constants E, G, K and v

It will be shown in Chapter 7 that the relationships between the elastic constants are given by

$$G = E/\{2(1 + v)\} \tag{2.13}$$

$$K = E/\{3(1 - 2v)\} \tag{2.14}$$

Some typical values of G and K are given in Table 2.2.

Table 2.2.   Some values of G and K (N/m^2).

Material	Steel	Aluminium alloy	Copper
G	8×10^{10}	2.6×10^{10}	4.4×10^{10}
K	1.67×10^{11}	6.86×10^{10}	1.33×10^{11}

(K for water $= 2 \times 10^9$)

2.7.1   THREE-DIMENSIONAL STRESS

Stress is a tensor, which can be represented diagrammatically by Fig. 2.16, and by the stress of equation (2.15).

$$\sigma_{ij} = \begin{pmatrix} \sigma_x & \tau_{xy} & \tau_{xz} \\ \tau_{yx} & \sigma_y & \tau_{yz} \\ \tau_{zx} & \tau_{zy} & \sigma_z \end{pmatrix} \tag{2.15}$$

where,

σ_{ij} = stress tensor of the 2nd order, using Einstein's notation,

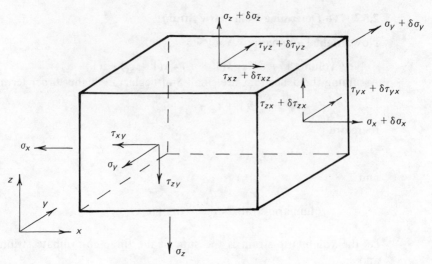

Fig. 2.16.   Three-dimensional stress system.

and

$$\tau_{xy} = \tau_{yx}, \text{ etc.}$$

If the front, bottom left corner of the cubic element of Fig. 2.16 is sliced off, to yield the tetrahedral sub-element of Fig. 2.17, it can be seen that equilibrium is achieved by the stress tensor σ_{ij} acting at an angle to the generic plane "abc". It is evident, therefore, that even for a tensile specimen undergoing

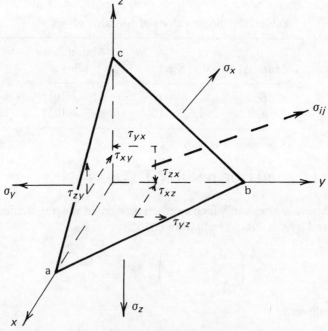

Fig. 2.17.   Three-dimensional stress.

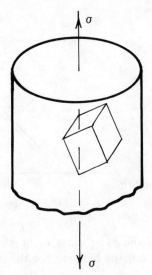

Fig. 2.18.   Elemental cube.

uniaxial stress, a three-dimensional system of stress can be obtained by examining an elemental cube, tilted at angle to the axis, as shown in Fig. 2.18.

2.8.1   THERMAL STRAIN

If a bar of length l and coefficient of linear expansion α is subjected to a temperature rise T, its length will increase by an amount $\alpha l T$, as shown in Fig. 2.19. Thus, at this temperature, the natural length of the bar is $l(1 + \alpha T)$. In this condition, although the bar has a thermal strain of αT, it has no thermal stress, but if the free expansion $\alpha l T$ is prevented from taking place, compressive thermal stresses will occur.

Such problems are of much importance in a number of practical situations, including railway lines and pipe systems, and the following two examples will be used to demonstrate the problem.

2.8.2   EXAMPLE 2.4   PROP UNDER TEMPERATURE CHANGE

A steel prop is used to stabilise a building, as shown in Fig. 2.20. If the compressive stress in the prop at a temperature of 20 C is 50 MN/m^2, what will the stress be in the prop if the temperature is raised to 30 C? At what

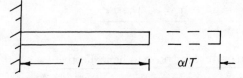

Fig. 2.19.   Free thermal expansion of a bar.

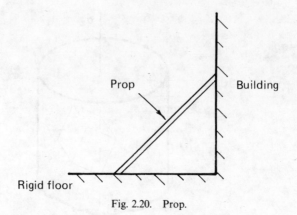

Fig. 2.20.   Prop.

temperature will the prop cease to be effective? It may be assumed that neither the floor nor the building nor the ends of the prop move.

$$E = 2 \times 10^{11} \text{ N/m}^2 \quad \alpha = 15 \times 10^{-6}/\text{C}$$

(a)   Additional thermal strain $= \dfrac{15 \times 10^{-6} \times l \times 10}{l}$

$$= -150 \times 10^{-6}$$

Thermal stress $\qquad = -2 \times 10^{11} \times 150 \times 10^{-6}$

$$= -30 \text{ MN/m}^2$$

Stress at 30 C $= -30 - 50 = -80 \text{ MN/m}^2$.

(b)   For the prop to be ineffective, it will be necessary for the temperature to drop, so that the initial compressive stress of 50 MN/m² is nullified, i.e.

Thermal stress $= 50 \text{ MN/m}^2$

Thermal strain $= \dfrac{50 \times 10^6}{2 \times 10^{11}} = 2.5 \times 10^{-4}$

or,

$$2.5 \times 10^{-4} = -\alpha T$$

where,

$$T = \text{temperature fall}$$

therefore

$$T = -16.67 \text{ C}$$

i.e.

temperature for prop to be ineffective

$$= 20 \text{ C} - 16.67 = 3.33 \text{ C}$$

2.8.3   EXAMPLE 2.5   STEEL RAIL UNDER TEMPERATURE CHANGE

A steel rail may be assumed to be in a stress-free condition at 10 C. If the stress required to cause buckling is -75 MN/m^2, what temperature rise will cause the rail to buckle, assuming that the rail is rigidly restrained at its ends, and that its material properties are as in Example 2.4?

$$\varepsilon = \frac{\sigma}{E} = \alpha T$$

or,

$$\frac{75 \times 10^6}{2 \times 10^{11}} = 15 \times 10^{-6}\, T$$

therefore

temperature rise $= T = 25$ C

and,

Temperature to cause buckling $= 35$ C

2.9.1   COMPOUND BARS

Compound bars are of much importance in a number of different branches of engineering, including reinforced concrete pillars, bimettalic bars, etc., and in this section, solution of such problems usually involves two important considerations, namely

(a)   compatibility (or consideration of displacements)
(b)   equilibrium

N.B.   It is necessary to introduce compatibility in this section as *compound bars are*, in gneral, *statically indeterminate*. To demonstrate the method of solution, the following two examples will be considered.

2.9.2   EXAMPLE 2.6   COMPOUND BAR UNDER TEMPERATURE CHANGE

A solid bar of cross-sectional area A_1, elastic modulus E_1 and coefficient of linear expansion α_1 is surrounded co-axially by a hollow tube of cross-sectional area A_2, elastic modulus E_2 and coefficient of linear expansion α_2, as shown in Fig. 2.21. If the two bars are secured firmly to each other, so that no slipping takes place with temperature change, determine the thermal stresses due to a temperature rise T. Both bars have an initial length l.

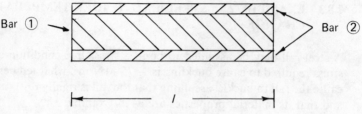

Fig. 2.21. Compound bar.

There are two unknowns; therefore two simultaneous equations will be required. The first equation can be obtained by considering the *compatibility* (i.e. "deflections") of the bars, with the aid of Fig. 2.22.

Free expansion of bar ① $= \alpha_1 \, lT$

Free expansion of bar ② $= \alpha_2 \, lT$

In practice, however, the final resting position of the compound bar will be *somewhere* between these two positions (i.e. at the position A–A). To achieve this, it will be necessary for bar ② to be pulled out by a distance $\varepsilon_2 l$ and for bar ② to be pushed in by a distance $\varepsilon_1 l$, where,

$\varepsilon_1 =$ compressive strain in ①

$\varepsilon_2 =$ tensile strain in ②

From considerations of compatibility ("deflections") in Fig. 2.22,

$$\alpha_1 lT - \varepsilon_1 l = \alpha_2 lT + \varepsilon_2 l$$

or,

$$\varepsilon_1 = (\alpha_1 - \alpha_2)T - \varepsilon_2$$

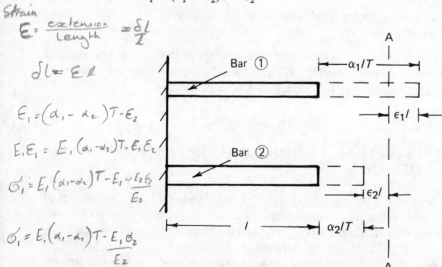

Fig. 2.22. "Deflections" of compound bar.

(handwritten annotations:)

Strain

$E = \dfrac{\text{extension}}{\text{length}} = \dfrac{\delta l}{l}$

$\delta l = E \, l$

$\varepsilon_1 = (\alpha_1 - \alpha_2)T - \varepsilon_2$

$E_1 \varepsilon_1 = E_1 (\alpha_1 - \alpha_2)T - E_1 \varepsilon_2$

$\sigma_1 = E_1 (\alpha_1 - \alpha_2)T - E_1 \dfrac{\varepsilon_2 \varepsilon_1}{E_2}$

$\sigma_1 = \dfrac{E_1 (\alpha_1 - \alpha_2)T - E_1 \sigma_2}{E_2}$

$-(2 \cdot 16)$

$\sigma_1 \text{, also} = 2.17$

or,

$$\sigma_1 = (\alpha_1 - \alpha_2)E_1 T - \sigma_2 E_1 / E_2 \tag{2.16}$$

To obtain the second simultaneous equation, it will be necessary to consider *equilibrium*.

Let,

$$F_1 = \text{compressive force in bar } \textcircled{1}$$

$$F_2 = \text{tensile force in bar } \textcircled{2}$$

Now,

$$F_1 = F_2$$

or

$$\sigma_1 A_1 = \sigma_2 A_2$$

therefore

$$\sigma_1 = \sigma_2 A_2 / A_1 \tag{2.17}$$

Equating (2.16) and (2.17):

$$\sigma_2 A_2 / A_1 = (\alpha_1 - \alpha_2)E_1 T - \sigma_2 E_1 / E_2$$

therefore

$$\sigma_2 = \frac{(\alpha_1 - \alpha_2)E_1 T}{(E_1/E_2 + A_2/A_1)}$$

or,

$$\sigma_2 = \frac{(\alpha_1 - \alpha_2)E_1 E_2 A_1 T}{(A_1 E_1 + A_2 E_2)} \text{ (tensile)} \tag{2.18}$$

and,

$$\sigma_1 = \frac{(\alpha_1 - \alpha_2)E_1 E_2 A_2 T}{(A_1 E_1 + A_2 E_2)} \text{ (compressive)} \tag{2.19}$$

2.9.3   EXAMPLE 2.7   COMPOUND BAR UNDER AXIAL LOAD

If the solid bar of Example 2.6 did not suffer temperature change, but instead was subjected to a tensile axial force P, as shown in Fig. 2.23, determine σ_1 and σ_2.

There are two unknowns; therefore two simultaneous equations will be required.

The first of these simultaneous equations can be obtained by considering *compatibility*, i.e.

deflection of bar $\textcircled{1}$ = deflection of bar $\textcircled{2}$

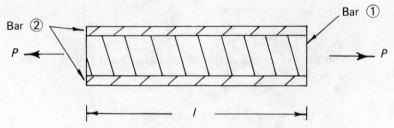

Fig. 2.23.   Compound bar under axial tension.

or,

$$\varepsilon_1 l = \varepsilon_2 l$$

$$\frac{\sigma_1}{E_1} = \frac{\sigma_2}{E_2}$$

therefore

$$\sigma_1 = \sigma_2 E_1/E_2 \qquad (2.20)$$

The second simultaneous equation can be obtained by considering *equilibrium*. Let,

$$F_1 = \text{tensile force in bar } ①$$

$$F_2 = \text{tensile force in bar } ②$$

Now,

$$P = F_1 + F_2$$

$$= \sigma_1 A_1 + \sigma_2 A_2 \qquad (2.21)$$

Substituting (2.20) into (2.21):

$$\sigma_2 = \frac{PE_2}{(A_1 E_1 + A_2 E_2)} \qquad (2.22)$$

and,

$$\sigma_1 = \frac{PE_1}{(A_1 E_1 + A_2 E_2)} \qquad (2.23)$$

N.B. If P is a compressive force, then both σ_1 and σ_2 will be compressive stresses.

2.9.4   EXAMPLE 2.8   CONCRETE PILLAR UNDER AXIAL LOAD

A concrete pillar, which is reinforced with steel rods, supports a compressive axial load of 1 MN. Determine σ_1 and σ_2, given the following:

$$\left.\begin{array}{l} A_1 = 3 \times 10^{-3} \text{ m}^2 \\ E_1 = 2 \times 10^{11} \text{ N/m}^2 \end{array}\right\} \text{Steel}$$

$$\left.\begin{array}{l} A_2 = 0.1 \text{ m}^2 \\ E_2 = 2 \times 10^{10} \text{ N/m}^2 \end{array}\right\} \text{Concrete}$$

What percentage of the total load does the steel reinforcement take?

(a) From (2.23):

$$\sigma_1 = -\frac{1 \times 10^6 \times 2 \times 10^{11}}{(6 \times 10^8 + 2 \times 10^9)} = \underline{-76.92 \text{ MN/m}^2} \qquad (2.24)$$

From (2.22):

$$\sigma_2 = -\frac{1 \times 10^6 \times 2 \times 10^{10}}{2.6 \times 10^9} = \underline{-7.69 \text{ MN/m}^2} \qquad (2.25)$$

(b) $F_1 = -76.92 \times 10^6 \times 3 \times 10^{-3} = \underline{2.308 \times 10^5 \text{ N}}$

Therefore percentage total load taken by the steel reinforcement
$= \underline{23.08}$

2.9.5   EXAMPLE 2.9   CONCRETE PILLAR UNDER AXIAL LOAD AND TEMPERATURE CHANGE

If the pillar of Example 2.8 were subjected to a temperature rise of 30 C, what would be the values of σ_1 and σ_2?

$\alpha_1 = 15 \times 10^{-6}/\text{C (steel)}$

$\alpha_2 = 12 \times 10^{-6}/\text{C (concrete)}$

As α_1 is larger than α_2, the effect of a temperature rise will cause the "thermal stresses" in ① to be compressive and those in ② to be tensile.
From (2.19):

σ_1 (thermal)

$$= -\frac{(15 \times 10^{-6} - 12 \times 10^{-6}) \times 2 \times 10^{11} \times 2 \times 10^{10} \times 0.1 \times 30}{(2.6 \times 10^9)}$$

$$= \underline{-13.85 \text{ MN/m}^2} \qquad (2.26)$$

From (2.18):

$$\sigma_2 \text{ (thermal)} = \underline{0.42 \text{ MN/m}^2} \qquad (2.27)$$

From (2.24) to (2.27):

$$\sigma_1 = -76.92 - 13.85 = \underline{-90.77 \text{ MN/m}^2}$$

$$\sigma_2 = -7.69 + 0.42 = \underline{-7.27 \text{ MN/m}^2}$$

2.9.6   EXAMPLE 2.10   WIRES UNDER TEMPERATURE CHANGE

A rigid horizontal bar is supported by three wires, where the outer wires are made from aluminium alloy and the middle wire from steel, as shown in Fig. 2.24. If the temperature of all three wires is raised by 50 C, what will be the thermal stresses in the wires?

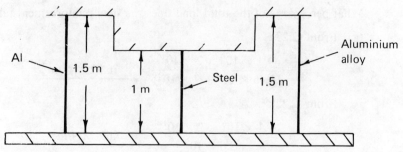

Fig. 2.24. Compound bar.

The following may be assumed:

$A_a = 3 \times 10^{-3}$ m² = sectional area of one aluminium wire
$E_a = 7 \times 10^{10}$ N/m² = elastic modulus of aluminium
$\alpha_a = 25 \times 10^{-6}$/C = coefficient of linear expansion of
　　　　　　　　　　　　aluminium
$A_s = 2 \times 10^{-3}$ m² = sectional area of steel wire
$E_s = 2 \times 10^{11}$ N/m² = elastic modulus of steel
$\alpha_s = 15 \times 10^{-6}$/C = coefficient of linear expansion of steel

Free expansion of aluminium $= 25 \times 10^{-6} \times 1.5 \times 50$

αₐ × lₐ × T

$$= 1.875 \times 10^{-3}$$

Free expansion of steel　　　　　$= 15 \times 10^{-6} \times 1 \times 50$

αₐ × lₛ × T

$$= 7.5 \times 10^{-4}$$

i.e. as the free expansion of the aluminium is greater than that of steel, the aluminium will be in compression and the steel in tension, owing to a temperature rise.

Let,

ε_a = compressive strain in aluminium

ε_s = tensile strain in steel

Now there are two unknowns; therefore two simultaneous equations will be required. The first of these can be obtained by considering *compatibility*, with the aid of Fig. 2.25. From Fig. 2.25, the final resting place of the compound bar will be at the position A–A, so that,

$$\alpha_a l_a T - \varepsilon_a l_a = \alpha_s l_s T + \varepsilon_s l_s$$

or,

$$\varepsilon_a = (\alpha_a l_a - \alpha_s l_s)T/l_a - \varepsilon_s l_s/l_a$$

× Eₐ　　　　　　　⇒ Eₐ εₐ = $\frac{E_a}{l_a}\left(\alpha_a l_a - \alpha_s l_s \right) T - E_a \times \left(\frac{E_s \, \varepsilon_s}{E_s}\right) \frac{l_s}{l_a}$　　／σₛ

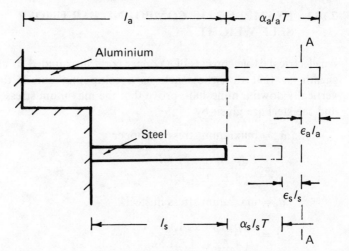

Fig. 2.25.

or,

$$\sigma_a = \frac{E_a}{l_a}\left(\alpha_a l_a - \alpha_s l_s\right)T - \frac{E_a}{E_s}\frac{l_s}{l_a}\sigma_s$$

$$\sigma_a = E_a\{\alpha_a l_a - \alpha_s l_s)T - \sigma_s l_s/E_s\}/l_a$$

$$= 7 \times 10^{10}\ (1.125 \times 10^{-3} - 5 \times 10^{-12}\sigma_s)/1.5$$

$$\sigma_a = (7.875 \times 10^7 - 0.35\ \sigma_s)/1.5$$

$$\sigma_a = 5.25 \times 10^7 - 0.233\ \sigma_s \qquad\qquad (2.28)$$

The second equation can be obtained from *equilibrium* considerations, where,

Tensile force in steel (F_s) = compressive force in aluminium (F_a)

or,

$$F_s = F_a$$

$$\sigma_s A_s = 2\sigma_a A_a \quad \leftarrow \text{2 aluminium wires}$$

therefore

$$\sigma_a = 0.333\ \sigma_s \qquad\qquad (2.29)$$

Equating (2.28) and (2.29):

$$5.25 \times 10^7 - 0.233\ \sigma_s = 0.333\ \sigma_s$$

therefore

$$\sigma_s = 92.76\ \text{MN/m}^2\ \text{(tensile)}$$

and,

$$\sigma_a = 30.9\ \text{MN/m}^2\ \text{(compressive)}$$

2.10.1 EXAMPLE 2.11 COMPOUND BAR UNDER SELF-WEIGHT

An electrical cable consists of a copper core, surrounded co-axially by a steel sheath, so that the whole acts as a compound bar. If this cable hangs vertically down a mineshaft, prove that the maximum stresses in the copper and the steel are given by

$$\hat{\sigma}_c = \text{maximum stress in copper}$$

$$= \frac{E_c(\rho_c A_c + \rho_s A_s)gl}{(A_c E_c + A_s E_s)}$$

$$\hat{\sigma}_s = \text{maximum stress in steel}$$

$$= \frac{E_s(\rho_c A_c + \rho_s A_s)gl}{(A_c E_c + A_s E_s)}$$

where,

ρ_c = density of copper
ρ_s = density of steel
E_c = elastic modulus of copper
E_s = elastic modulus of steel
g = acceleration due to gravity
A_c = cross-sectional area of copper
A_s = cross-sectional area of steel
l = length of cable

2.10.2 Compatibility Considerations

Let,

δ_c = maximum deflection of the copper core

δ_s = maximum deflection of the steel sheath (δ_s)

Now,

$$\delta_c = \delta_s$$

or,

$$l\varepsilon_c = l\varepsilon_s$$

where,

ε_c = maximum strain in copper

ε_s = maximum strain in steel

or,

$$\frac{\hat{\sigma}_c}{E_c} = \frac{\hat{\sigma}_s}{E_s}$$

i.e.

$$\hat{\sigma}_c = \hat{\sigma}_s E_c/E_s \tag{2.30}$$

2.10.3   Equilibrium Considerations

Weight of cable $= (\rho_c A_c + \rho_s A_s)gl$
Resolving vertically

$$(\rho_c A_c + \rho_s A_s)gl = \hat{\sigma}_c * A_c + \hat{\sigma}_s * A_s \tag{2.31}$$

Substituting equation (2.30) into (2.31):

$$(\rho_c A_c + \rho_s A_s)gl = \hat{\sigma}_s A_c E_c/E_s + \hat{\sigma}_s A_s$$

$$\hat{\sigma}_s \left(\frac{A_c E_c}{E_s} + A_s \right) = (\rho_c A_c + \rho_s A_s)gl$$

$$\hat{\sigma}_s = \frac{(\rho_c A_c + \rho_s A_s)gl}{\left(\dfrac{A_c E_c}{E_s} + A_s \right)}$$

$$\hat{\sigma}_s = \frac{E_s(\rho_c A_c + \rho_s A_s)gl}{(A_c E_c + A_s E_s)} \tag{2.32}$$

Substituting equation (2.32) into (2.30):

$$\hat{\sigma}_c = \frac{E_c(\rho_c A_c + \rho_s A_s)gl}{(A_c E_c + A_s E_s)} \tag{2.33}$$

2.11.1   EXAMPLE 2.12   BOLT/TUBE COMPOUND BAR

A compound bar consists of a steel bolt surround co-axially by an aluminium alloy tube, as shown in Fig. 2.26. Assuming that the nut on the end of the steel bolt is initially "just" hand-tight, determine the strains in the bolt and the tube, if the nut is rotated clockwise by an angle θ, relative to the other end.

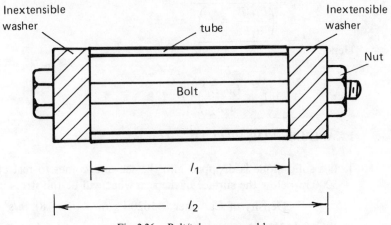

Fig. 2.26.   Bolt/tube compound bar.

Let,

ε_1 = strain in the aluminium alloy tube = δ_1/l_1
ε_2 = strain in the steel bolt = δ_2/l_2
a_1 = sectional area of aluminium alloy
a_2 = sectional area of steel
l_1 = length of aluminium alloy tube
l_2 = length of steel bolt
$\delta = \theta \times$ pitch of thread/360
θ = rotation in degrees
E_1 = elastic modulus for aluminium alloy
E_2 = elastic modulus for steel

Now, when the nut is turned clockwise, by an angle θ, the aluminium alloy tube will decrease its length by δ_1, and the steel bolt will increase its length by δ_2, where

$$\delta = \delta_1 + \delta_2$$

$$\delta = \varepsilon_1 l_1 + \varepsilon_2 l_2 \tag{2.34}$$

From equilibrium considerations,

$$\sigma_1 a_1 = \sigma_2 a_2$$

or

$$E_1 \varepsilon_1 a_1 = E_2 \varepsilon_2 a_2$$

i.e.

$$\varepsilon_1 = \frac{\varepsilon_2 a_2 E_2}{a_1 E_1} \tag{2.35}$$

Substituting equation (2.35) into (2.34),

$$\varepsilon_2 \left(\frac{l_1 a_2 E_2}{a_1 E_1} + l_2 \right) = \delta$$

$$\varepsilon_2 = \frac{\delta}{\left(\dfrac{l_1 a_2 E_2}{a_1 E_1} + l_2 \right)} \tag{2.36}$$

Hence,

$$\varepsilon_1 = \frac{-\delta}{\left(l_1 + \dfrac{l_2 a_1 E_1}{a_2 E_2} \right)} \tag{2.37}$$

EXAMPLES FOR PRACTICE 2

1. If a solid stone is dropped into the sea and comes to rest at a depth of 5000 m below the surface of the sea, what will be the stress in the stone?

    Density of sea water = 1020 kg/m^3 $g = 9.81$ m/s^2

    $\{ -50 \text{ MN/m}^2 \}$

2. A solid bar of length 1 m consists of three shorter sections firmly joined together.

    Assuming the following apply, determine the change in length of the bar when it is subjected to an axial pull of 50 kN?

$$E = 2 \times 10^{11} \text{ N/m}^2$$

Section	Length (m)	Diameter (mm)
1	0.2	15
2	0.3	20
3	0.5	30

{0.699 mm}

3. If the bar of Example 2 were made from three different materials, with the following elastic moduli, determine the change in length of the bar:

Section	$E(\text{N/m}^2)$
1	2×10^{11}
2	7×10^{10}
3	1×10^{11}

{1.319 mm}

4. A circular section solid bar of linear taper is subjected to an axial pull of 0.1 MN, as shown in Fig. Q.2.4. If $E = 2 \times 10^{11}$ N/m², by how much will the bar extend?

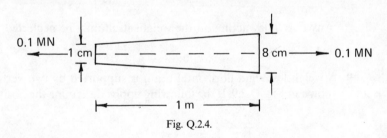

Fig. Q.2.4.

{0.796 mm}

5. If the bar of Example 4 were prevented from moving axially by two rigid walls and subjected to a temperature rise of 10 C, what would be the maximum stress in the bar? Assume the 0.1 MN load is not acting.

$$\alpha = 15 \times 10^{-6}/\text{C}$$

{ − 240 MN/m² at the smaller end}

6. An electrical cable consists of a copper core, surrounded co-axially by a steel sheath, so that the two can be assumed to act as a compound bar. If

the cable hangs down a vertical mineshaft, determine the maximum permissible length of the cable, assuming the following apply:

$$A_c = 1 \times 10^{-4} \text{ m}^2 \quad = \text{sectional area of copper}$$
$$E_c = 1 \times 10^{11} \text{ N/m}^2 = \text{elastic modulus of copper}$$
$$\rho_c = 8960 \text{ kg/m}^3 \quad = \text{density of copper}$$

Maximum permissible stress in copper = 30 MN/m^2

$$A_s = 0.2 \times 10^{-4} \text{ m}^2 = \text{sectional area of steel}$$
$$E_s = 2 \times 10^{11} \text{ N/m}^2 = \text{elastic modulus of steel}$$
$$\rho_s = 7860 \text{ kg/m}^3 \quad = \text{density of steel}$$

Maximum permissible stress in steel = 100 MN/m^2

$$g = 9.81 \text{ m/s}^2$$

{406.5 m}

7. How much will the cable of Example 6 stretch, due to self-weight?

{61 mm}

8. If a weight of 100 kN were lowered into the sea, via a steel cable of cross-sectional area 8×10^{-4} m^2, what would be the maximum permissible depth that the weight could be lowered if the following apply:

Density of steel = 7860 kg/m^3
Density of sea water = 1020 kg/m^3

Maximum permissible stress in steel = 200 MN/m^2
$$g = 9.81 \text{ m/s}^2$$

Any buoyancy acting on the weight itself may be neglected.

{1118 m}

9. A weightless rigid horizontal beam is supported by two vertical wires, as shown in Fig. Q.2.9. If the following apply, determine the position from the

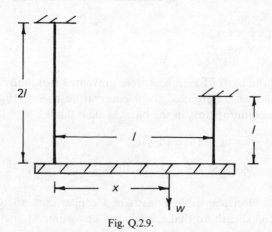

Fig. Q.2.9.

left that a weight W can be suspended, so that the bar will remain horizontal when the wires stretch.

left wire

$$\text{cross-sectional area} = 2A$$
$$\text{elastic modulus} = E$$
$$\text{length} = 2l$$

right wire

$$\text{cross-sectional area} = A$$
$$\text{elastic modulus} = 3E$$
$$\text{length} = l$$

$\{0.75 \, l\}$

3

Geometrical Properties of Symmetrical Sections

3.1.1

The geometrical properties of sections is of much importance in a number of different branches of engineering, including stress analysis. For example, if a beam is subjected to bending action, its bending stiffness will depend not only on the material properties of the beam, but also on the geometrical properties of its cross-section. Some typical cross-sections for symmetrical sections are shown in Fig. 3.1, where it is evident that, providing the material properties of the section are the same, the bending resistances of the sections are dependent on their geometrical properties.

3.1.2

After many years of experience, structural engineers have found that if beams are made from steel or aluminium alloy, the cross-sections of Figs. 3.1(d) and 3.1(e) usually provide a better strength: weight ratio than do the sections of Figs 3.1(a) to 3.1(c).

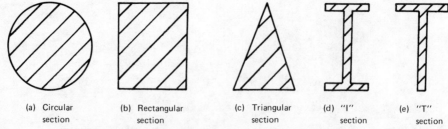

(a) Circular section   (b) Rectangular section   (c) Triangular section   (d) "I" section   (e) "T" section

Fig. 3.1.   Some symmetrical cross-sections of beams.

The section of Fig. 3.1(d) is known as a *rolled steel joist* (RSJ) and that of Fig. 3.1(e) is known as a *tee section*.

3.2.1   CENTROID

This is the centre of the *moment of area* of a plane figure; or if the plane figure is of uniform thickness, this is the same position as the centre of gravity. This position is of much importance in elastic stress analysis.

At the centroid, the following equations apply:

$$\int y \, dA = 0 \quad \text{and} \quad \int x \, dA = 0 \tag{3.1}$$

where x is the horizontal axis, and y is the vertical axis, mutually perpendicular to x, as shown in Fig. 3.2.

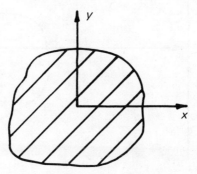

Fig. 3.2.   Plane figure.

3.2.2   Centroidal Axes

These are lines that pass through the centroid.

3.2.3   Centre of Area

For a plane figure, this is obtained from the following considerations:

Area above the horizontal central axis = area below the
horizontal central axis   (3.2)

Area to the left of the vertical central axis = area to
the right of the vertical central axis   (3.3)

3.2.4   Central Axes

These are lines that pass through the centre of area.

3.2.5 SECOND MOMENT OF AREA

The second moment of area of the section of Fig. 3.3 about $XX =$

$$I_{XX} = \int y^2 \, dA \tag{3.4}$$

The second moment of area of the section of Fig. 3.3 about $YY =$

$$I_{YY} = \int x^2 \, dA \tag{3.5}$$

where,

$$\int dA = \text{area of section}$$

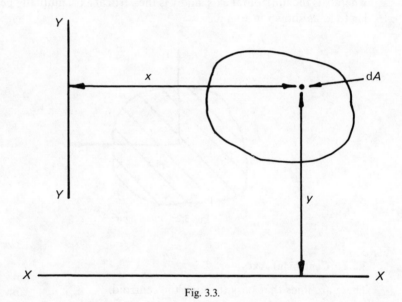

Fig. 3.3.

3.2.6 POLAR SECOND MOMENT OF AREA

The polar second moment of area of the circular section of Fig. 3.4, about its centre "O", is given by

$$J = \int_0^R r^2 \, dA \tag{3.6}$$

where

$$dA = 2\pi r . dr$$

J is of importance in the torsion of circular sections.

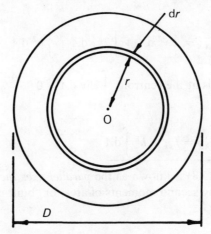

Fig. 3.4.   Circular section.

3.3.1 PARALLEL AXES THEOREM

Consider the plane section of Fig. 3.5, which has a second moment of area about its centroid equal to I_{XX}, and suppose that it is required to determine the second moment of area about XX, where XX is parallel to xx and that the perpendicular distance between the two axes is h. Now,

$$I_{xx} = \int y^2 \, \mathrm{d}A$$

and,

$$I_{XX} = \int (y + h)^2 \, \mathrm{d}A$$

$$= \int (y^2 + 2hy + h^2) \, \mathrm{d}A$$

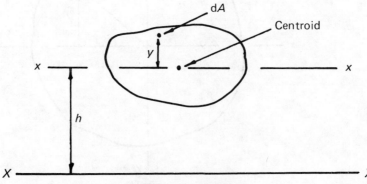

Fig. 3.5.   Parallel axes.

or,

$$I_{XX} = \int y^2 \, dA + \int 2hy \, dA + \int h^2 \, dA \tag{3.7}$$

but as xx is at the centroid, $\int 2hy \, dA = 0$

Therefore,

$$I_{XX} = I_{xx} + h^2 \int dA \tag{3.8}$$

Equation (3.8) is known as the *parallel axes theorem* and it is important in determining second moments of area for "built-up" sections, such as R.S.J.s, tees, etc.

3.3.2 PERPENDICULAR AXES THEOREM

From Fig. 3.6, it can be seen that

$$I_{xx} = \int y^2 \, dA$$

$$I_{yy} = \int x^2 \, dA$$

Now from equation (3.6)

$$J = \int r^2 \, dA$$

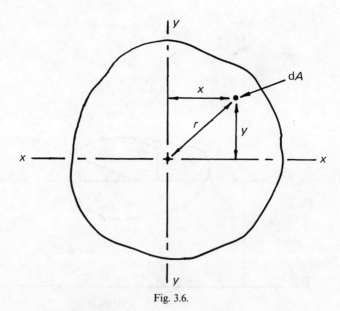

Fig. 3.6.

but as $x^2 + y^2 = r^2$,

$$J = I_{xx} + I_{yy} \tag{3.9}$$

Equation (3.9) is known as the *perpendicular axes theorem*, which states that the sum of the second moments of area of two mutually perpendicular axes is equal to the polar second moment of area about the point where these two axes cross.

3.3.3

To demonstrate the theories described in this section, the following examples will be considered.

3.4.1   EXAMPLE 3.1   TRIANGLE

Determine the positions of the centroidal and central axes for the isosceles triangle of Fig. 3.7. Hence, or otherwise, determine the second moments of area about the centroid and the base XX. Verify the parallel axes theorem by induction.

3.4.2   To Find Area

$$\int dA = \int b \, dy$$

but,

$$b = B(1 - y/H)$$

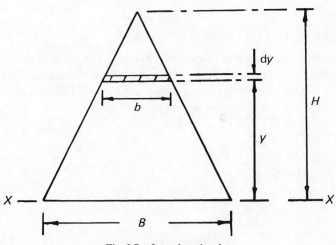

Fig. 3.7.   Isosceles triangle.

therefore,

$$\int b \, dy = B \int_0^H (1 - y/H) dy$$

$$= B \left[y - y^2/2H \right]_0^H$$

or area

$$= A = BH/2 \tag{3.10}$$

3.4.3   To Find Centroidal Axis

Let $\bar{y}$ = distance of centroid from XX therefore 1st moment of area about $XX = A\bar{y}$

$$= \int_0^H y.b \, dy$$

$$= B \int_0^H (y - y^2/H) dy$$

$$= B \left[y^2/2 - y^3/3H \right]_0^H$$

or,

$$A\bar{y} = BH^2/6$$

therefore,

$$\bar{y} = H/3 \tag{3.11}$$

i.e. the distance of the centroid above $XX = H/3$.

3.4.4   To Find Central Axis

Let $\bar{Y}$ = distance of central axis above XX. Consider the isosceles triangle of Fig. 3.8 and equate areas above and below the central axis.

Area above the central axis = area below the central axis.

$$\frac{b(H - \bar{Y})}{2} = \frac{(B + b)\bar{Y}}{2}$$

but $b = B(1 - \bar{Y}/H)$.

Therefore,

$$B(H - \bar{Y} - \bar{Y} + \bar{Y}^2/H) = B\bar{Y} + B(\bar{Y} - \bar{Y}^2/H)$$

or,

$$2\bar{Y}^2 - 4\bar{Y}H + H^2 = 0$$

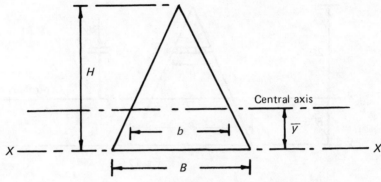

Fig. 3.8.   Central axis.

therefore,

$$\bar{Y} = \frac{4H \pm \sqrt{(16 - 8)}\,H}{4}$$

$$\bar{Y} = 0.293H \tag{3.12}$$

i.e. for this case, the central axis is 12.1% below the centroidal axis.

In stress analysis, the position of the central axis is only of importance in plasticity; so for the remainder of the present chapter, we will restrict our interest to the centroidal axis, which is of much interest in elastic theory.

3.4.5   To Find I_{xx}

I_{xx} = second moment of area about the centroidal axis, as in Fig. 3.9.

Now,

$$b = 2B/3 - By/H$$

and,

$$I_{xx} = \int_{-H/3}^{2H/3} y^2\, b\mathrm{d}y$$

$$= B \int_{-H/3}^{2H/3} \left(\frac{2y^2}{3} - \frac{y^3}{H}\right) \mathrm{d}y$$

$$= B \left[\frac{2y^3}{9} - \frac{y^4}{4H}\right]_{-H/3}^{2H/3}$$

$$= BH^3 \left\{\left[\frac{16}{27 \times 9} - \frac{16}{4 \times 81}\right] - \left[\frac{-2}{9 \times 27} - \frac{1}{4 \times 81}\right]\right\}$$

$$\underline{I_{xx} = BH^3/36} \tag{3.13}$$

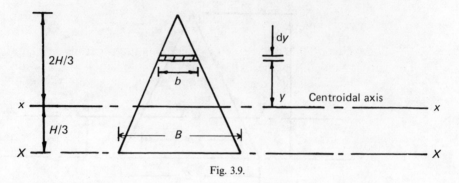

Fig. 3.9.

Similarly, from Fig. 3.10,

$$I_{XX} = \int_0^H y^2 \, b \mathrm{d}y$$

but,

$$b = B(1 - y/H)$$

$$I_{XX} = B \int_0^H (y^2 - y^3/H)\mathrm{d}y$$

$$= B \left[\frac{y^3}{3} - \frac{y^4}{4H} \right]_0^H$$

$$\underline{I_{XX} = BH^3/12} \tag{3.14}$$

3.4.6 Check on Parallel Axes Theorem

From equation (3.8),

$$I_{XX} = I_{xx} + \frac{BH}{2} * \left(\frac{H}{3} \right)^2$$

$$= \frac{BH^3}{12} \text{ (as required)}$$

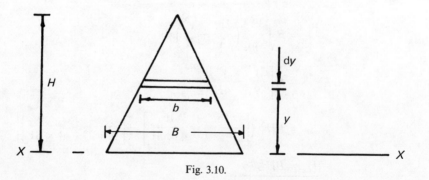

Fig. 3.10.

3.5.1   EXAMPLE 3.2   ELLIPSE

Determine the area and second moments of area about the major and minor axes for the elliptical section of Fig. 3.11.

Now the equation of an ellipse is

$$\frac{x^2}{a^2} + \frac{y^2}{b^2} = 1$$

therefore

$$y^2 = (1 - x^2/a^2)b^2$$

Let

$$x = a \cos \phi \tag{3.15}$$

therefore

$$y = (1 - a^2 \cos^2 \phi/a^2)^{1/2}b$$

$$y = b \sin \phi \tag{3.16}$$

$$\mathrm{d}y = b \cos \phi \, \mathrm{d}\phi \tag{3.17}$$

Now,

$$A = \text{area of elliptical figure}$$

$$= 4 \int x \, \mathrm{d}y = 4 \int_0^{\pi/2} a \cos \phi \, b \cos \phi \, \mathrm{d}\phi$$

$$= 4ab \int_0^{\pi/2} \cos^2 \phi \, \mathrm{d}\phi$$

$$= 4ab \int_0^{\pi/2} \frac{(1 + \cos 2\phi)}{2} \, \mathrm{d}\phi$$

therefore

$$A = \pi ab \tag{3.18}$$

From Fig. 3.11, it can be seen that

$$I_{xx} = 4 \int y^2 x \, \mathrm{d}y$$

$$= 4 \int b^2 \sin^2 \phi \, a \cos \phi \, b \cos \phi \, \mathrm{d}\phi$$

$$= 4ab^3 \int \frac{(1 - \cos 2\phi)}{2} \frac{(1 + \cos 2\phi)}{2} \, \mathrm{d}\phi$$

$$= ab^3 \int (1 - \cos^2 2\phi) \, \mathrm{d}\phi$$

$$= ab^3 \int_0^{\pi/2} \left[1 - \frac{(1 + \cos 4\phi)}{2} \right] \mathrm{d}\phi$$

$$\underline{I_{xx} = \pi ab^3/4} \tag{3.19}$$

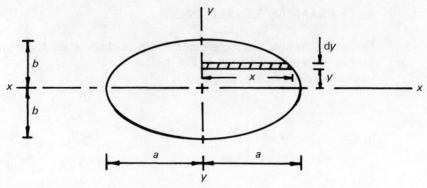

Fig. 3.11. Elliptical section.

Similarly, it can be proven that

$$I_{yy} = \pi a^3 b/4 \qquad (3.20)$$

For a circle of radius R (or diameter D),

$$R = a = b$$

therefore

$$I_{xx} = I_{yy} = \frac{\pi R^4}{4} = \frac{\pi D^4}{64} \qquad (3.21)$$

and,

$$J = \frac{\pi D^4}{32} \qquad (3.22)$$

3.6.1 EXAMPLE 3.3 RECTANGLE

Determine the second moments of area about x–x and X–X for the rectangle of Fig. 3.12, and verify the parallel axes theorem by induction.

$$I_{xx} = \int_{-D/2}^{D/2} y^2 B \, dy$$

$$= B \left[\frac{y^3}{3} \right]_{-D/2}^{D/2}$$

$$I_{xx} = BD^3/12 \qquad (3.23)$$

$$I_{XX} = B \int_{-D/2}^{D/2} (y + D/2)^2 \, dy$$

$$= B \int_{-D/2}^{D/2} (y^2 + Dy + D^2/4) \, dy$$

$$= B \left[y^3/3 + Dy^2/2 + D^2y/4 \right]_{-D/2}^{D/2}$$

$$I_{XX} = BD^3/3 \qquad (3.24)$$

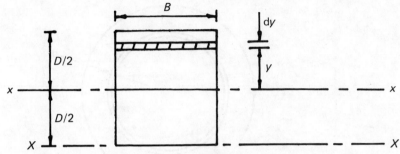

Fig. 3.12.   Rectangle.

3.6.2   Check on Parallel Axes Theorem

$$I_{XX} = I_{xx} + A(D/2)^2 = BD^3/3 \text{ (as required)}$$

3.6.3

For the parallelogram of Fig. 3.13, it can be proven that

$$I_{XX} = BD^3/3$$

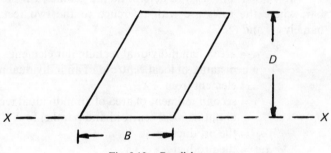

Fig. 3.13.   Parallelogram.

3.7.1   EXAMPLE 3.4   CIRCULAR SECTION

Determine the polar second moment of area for the circular section of Fig. 3.14.

$$J = \int_0^R r^2\, 2\pi r\, \mathrm{d}r$$

$$= 2\pi \left[\frac{r^4}{4} \right]_0^R$$

$$J = \frac{\pi R^4}{2} = \frac{\pi D^4}{32} \tag{3.25}$$

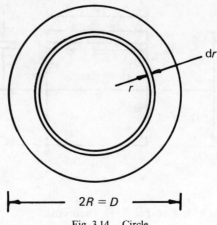

Fig. 3.14. Circle.

3.8.1 EXAMPLE 3.5 TEE SECTION

Determine the position of the centroidal axis xx and the second moment of area about this axis for the tee-bar of Fig. 3.15.

Table 3.1 will be used to determine the geometrical properties of the tee-bar, where the rows are with reference to the two rectangular elements, namely ① and ②.

a = area of an individual rectangular element.

y = distance of local centroid of an individual rectangular element from XX.

i = second moment of area of an individual rectangular element about its local centroid and parallel to XX

ay = the product $a * y$.

ay^2 = the product $a * y * y$.

$\sum$ = summation of the appropriate column.

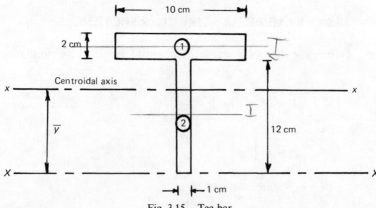

Fig. 3.15. Tee-bar.

Table 3.1.   Geometrical calculations for tee-bar

Section	$a(\text{m}^2)$	$y(\text{m})$	$ay(\text{m}^3)$	$ay^2(\text{m}^4)$	$i(\text{m}^4)$
①	2×10^{-3}	0.13	2.6×10^{-4}	3.38×10^{-5}	$\dfrac{0.1 \times 0.02^3}{12} = 6.67 \times 10^{-8}$
②	1.2×10^{-3}	0.06	7.2×10^{-5}	4.32×10^{-6}	$\dfrac{0.01 \times 0.12^3}{12} = 1.44 \times 10^{-6}$
	3.2×10^{-3}	—	3.32×10^{-4}	3.812×10^{-5}	1.51×10^{-6}

From Table 3.1,

$$\bar{y} = \sum ay / \sum a = 3.32 \times 10^{-4}/3.2 \times 10^{-3} \qquad (3.26)$$
$$\bar{y} = 0.104 \text{ m}$$

$$I_{XX} = \sum ay^2 + \sum i \qquad (3.27)$$
$$= 3.812 \times 10^{-5} + 1.51 \times 10^{-6}$$

$$I_{XX} = 3.963 \times 10^{-5} \text{ m}^4$$

$$I_{xx} = I_{XX} - (\bar{y})^2 \sum a \qquad (3.28)$$
$$= 3.963 \times 10^{-5} - (0.104)^2 \times 3.2 \times 10^{-3}$$

$$I_{xx} = 5.02 \times 10^{-6} \text{ m}^4$$

3.9.1 EXAMPLE 3.6 RECTANGLE WITH HOLE

Determine the position of the centroidal axis of the section of Fig. 3.16 and also the second moment of area about this axis.

THINK OF THIS HOLE AS BEING A NEGATIVE AREA SO IT IS SUBTRACTED

①rectangle
②=hole

① I = BD³/12 rectangle

② I = πd⁴/64 circle about diameter

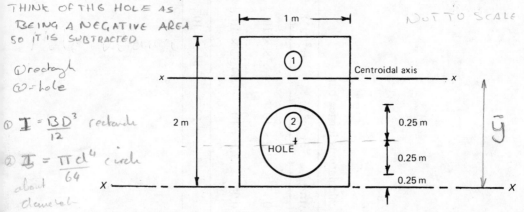

NOT TO SCALE

Fig. 3.16.

The determination of the geometrical properties of the section of Fig. 3.16 will be aided with the calculations of Table 3.2, where the symbols are defined in Section 3.8.1. From Table 3.2,

$$\bar{y} = \sum ay / \sum a = 1.054 \text{ m}$$
$$I_{XX} = \sum ay^2 + \sum i = 2.615 \text{ m}^4$$
$$I_{xx} = I_{XX} - (\bar{y})^2 \sum a = 0.611 \text{ m}^4$$

3.10.1 CALCULATION OF I THROUGH NUMERICAL INTEGRATION

If I is required for an arbitrarily shaped section, such as that shown in Fig. 3.17, the calculation for I can be carried out through numerical integration.

Table 3.2.   Geometrical calculations.

Section	$a(m^2)$	$y(m)$	$ay(m^3)$	$ay^2(m^4)$	$i(m^4)$
①	2	1	2	2	$1 \times 2^3/12 = 0.66667$
②	-0.1963	0.5	-0.098	-0.0491	$-\pi \times (5 \times 10^{-1})^4/64 = -3.07 \times 10^{-3}$
	1.8037	—	1.902	1.9509	0.664

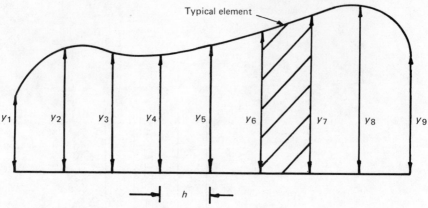

Fig. 3.17.   Arbitrarily shaped section.

If the numerical integration is based on a Simpson's rule approach, then the section must be divided into an equal number of elements, where the number of elements must be even.

3.10.2   Proof of Simpson's Rule

Simpon's rule is based on employing a parabola to describe the function over any three stations, as shown in Fig. 3.18. The equation is

$$y = a + bx + cx^2 \tag{3.29}$$

where a, b and c are arbitrary constants.

To obtain the three unknown constants, it will be necessary to obtain three simultaneous equations by putting in the *boundary values* for y in equation (3.29), as follows.

$$\underline{@\ x = 0, y = y_2 = a}$$

therefore

$$\underline{a = y_2} \tag{3.30}$$

or

$$\underline{@\ x = -h, y = y_1}$$

$$y_1 - y_2 = -bh + ch^2 \tag{3.31}$$

$$\underline{@\ x = h, y = y_3}$$

or

$$y_3 - y_2 = bh + ch^2 \tag{3.32}$$

Adding (3.31) and (3.32):

$$c = \frac{(y_1 - 2y_2 + y_3)}{2h^2} \tag{3.33}$$

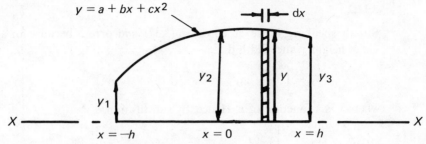

Fig. 3.18.   Parabolic variation.

Substituting equation (3.33) into (3.32):

$$b = \frac{(-y_1 + y_3)}{2h} \tag{3.34}$$

Now,

$$A = \text{area of section}$$

$$= \int_{-h}^{h} y \, dx$$

$$= \int_{-h}^{h} (a + bx + cx^2) \, dx$$

$$= \left(ah + \frac{ch^3}{3} \right) * 2 \tag{3.35}$$

Substituting equations (3.30) and (3.33) into (3.35)

$$A = 2h\{y_2 + \frac{1}{6}(y_1 - 2y_2 + y_3)\}$$

$$A = \frac{h}{3}(y_1 + 4y_2 + y_3) \tag{3.36}$$

Equation (3.36) is known as *Simpson's rule* for calculating areas.

3.11.1   Naval Architect's Method of Numerically Calculating I_{XX} for a Ship's Water Plane

The naval acrchitect's method of calculating I_{XX}, which is based on Simpson's rule, is given by equation (3.37). This expression is reasonable for gentle curves with relatively small values of h.

$$I_{XX} = \frac{h}{9}(y_1^3 + 4y_2^3 + y_3^3) \tag{3.37}$$

3.11.2

Strictly speaking, however, equation (3.37) is incorrect, because for a rectangle of height y and width dx,

$$I_{XX} = \int_{-h}^{h} \frac{y^3}{3} \, dx,$$

where y is a function of x. Hence, by substitution,

$$I_{XX} = \tfrac{1}{3} \int_{-h}^{h} (a + bx + cx^2)^3 \, dx,$$

which is very different to equation (3.37), i.e.

$$I_{XX} = \tfrac{2}{3} \left(a^3h + \frac{c^3h^7}{7} + ab^2h^3 + a^2\,ch^3 + \tfrac{3}{5}b^2ch^5 + \frac{3ac^2}{5} h^5 \right)$$

$$= \frac{2h}{3} \left\{ y_2^3 + \frac{1}{7 \times 8} (y_1^3 - 6y_1^2y_2 + 3y_1^2y_3 \right.$$

$$+ 12y_1y_2^2 - 12y_1y_2y_3 + 3y_1y_3^2 - 8y_2^3$$

$$+ 12y_2^2y_3 - 6y_2y_3^2 + y_3^3)$$

$$+ \frac{y_2}{4} (y_1^2 - 2y_1y_3 + y_3^2) + \frac{y_2^2}{2} (y_1 - 2y_2 + y_3)$$

$$+ \tfrac{3}{40}(y_1^2 - 2y_1y_3 + y_3^2) \cdot (y_1 - 2y_2 + y_3)$$

$$\left. + \tfrac{3}{20}y_2(y_1^2 - 4y_1y_2 + 2y_1y_3 + 4y_2^2 - 4y_2y_3 + y_3^2) \right\}$$

which, when rearranged, becomes

$$I_{XX} = \frac{2h}{3} \{ \tfrac{13}{140}(y_1^3 + y_3^3) + \tfrac{16}{35}y_2^3 + \tfrac{1}{7}(y_1^2 \, y_2 + y_2 y_3^2)$$

$$- \tfrac{3}{140}(y_1^2 y_3 + y_1 y_3^2) + \tfrac{4}{35}(y_1 y_2^2 + y_2^2 y_3) - \tfrac{4}{35}(y_1 y_2 y_3)\} \quad (3.38)$$

3.11.3   EXAMPLE 3.7   ARBITRARILY SHAPED FIGURE

Calculate I_{XX} for the section of Fig. 3.19 by equations (3.37) and (3.38), and then compare the two results.

From equation (3.37):

$$I_{XX} = \tfrac{1}{9}(0.8^3 + 4 \times 0.6^3 + 1.4^3) = 0.4578 \text{ m}^4$$

From equation (3.38):

$$I_{XX} = \tfrac{2}{3}\{ \tfrac{13}{140} \times 3.256 + \tfrac{16}{35} \times 0.216 + \tfrac{1}{7} \times 1.56 - \tfrac{3}{140} \times 2.464$$

$$+ \tfrac{4}{35} \times 0.792 - \tfrac{4}{35} \times 0.672 \}$$

$$\underline{I_{XX} = 0.39 \text{ m}^4}$$

i.e percentage difference = 17.41

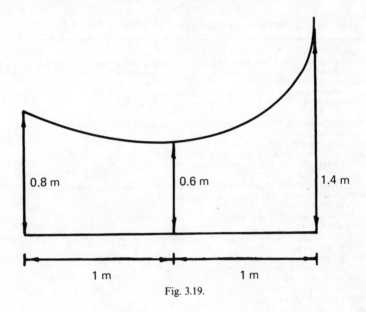

Fig. 3.19.

3.12.1   Computer Program for Calculating $\bar{y}$ and I_{xx}

Table 3.3 gives a listing of a computer program, in BASIC, for calculating $\bar{y}$ and I_{xx} for symmetrical sections, and Tables 3.4 and 3.5 give the outputs for this program for problems 2(b) and 2(d), respectively, from the Examples for Practice, at the end of this chapter.

It should be noted from Tables 3.4 and 3.5 that the units being used must not be mixed (i.e. do not use a mixture of centimetres with metres or millimetres, etc.).

3.12.2   Program Input

Input n—number of elements
FOR $i = 1$ TO n
Input $a(i)$—area of element i
Input $y(i)$—distance of centroid of element i from X-X
Input $iO(i)$—local second moment of area of element i
NEXT I

3.12.3   Program Output

area $= \sum A$
ybar $= \bar{y}$
ina $= I_{xx}$

Table 3.3.   Computer program for calculating $\bar{y}$ and I_{xx}.

```
100 CLS
110 REMark program for second moments of area for symmetrical sections
120 PRINT:PRINT"program for second moments of area for symmetrical sections"
130 PRINT:PRINT"copyright of Dr.C.T.F.ROSS"
140 PRINT:PRINT"type in the number of sections"
150 INPUT n
160 IF n>0 THEN GO TO 180
170 PRINT:PRINT"incorrect data":GO TO 140
180 absn=ABS(n)
190 in=INT(absn)
200 IF in=n THEN GO TO 220
210 PRINT:PRINT"incorrect data":GO TO 140
220 DIM a(n),y(n),i0(n)
230 PRINT:PRINT"type in the details of each element":PRINT
240 area=0:moment=0:second=0:ilocal=0
250 FOR i=1 TO n
260 PRINT"elemental area(";i;")=";:INPUT a(i)
270 PRINT"element centroid from XX _i.e y(";i;")=";:INPUT y(i)
280 PRINT"elemental local 2nd moment of area io(";i;")=";:INPUT i0(i)
290 area=area+a(i)
300 moment=moment+a(i)*y(i)
310 second=second+a(i)*y(i)*y(i)
320 ilocal=ilocal+i0(i)
330 NEXT i
340 ybar=moment/area
350 ixx=ilocal+second
360 ina=ixx-ybar^2*area
370 OPEN£3,ser1
380 PRINT£3,"number of elements=";n
390 FOR i=1 TO n
400 PRINT£3,"element area a(";i;")=";a(i)
410 PRINT£3,"element centroid y(";i;")=";y(i)
420 PRINT   "elemental local 2nd moment of area=";i0(i)
430 NEXT
440 PRINT£3
450 PRINT£3
460 PRINT£3,"sectional area=";area
470 PRINT£3,"section centroid (ybar) from XX=";ybar
480 PRINT£3,"2nd moment of area of section about its centroid=";ina
490 CLOSE£3
500 STOP
```

Table 3.4.   Computer output for problem 2(b).

```
number of elements=3
element area a(1)=8
element centroid y(1)=14.5
elemental local 2nd moment of area=.666667
element area a(2)=12
element centroid y(2)=8
elemental local 2nd moment of area=144
element area a(3)=20
element centroid y(3)=1
elemental local 2nd moment of area=6.6667


sectional area=40
section centroid (ybar) from XX=5.8
2nd moment of area of section about its centroid=1275.733
```

Table 3.5.   Computer output for problem 2(d).

```
number of elements=2
element area a (1) =154
element centroid y (1) =7
elemental local 2nd moment of area=2515.3
element area a (2) =-78.54
element centroid y (2) =6
elemental local 2nd moment of area=-490.9


sectional area=75.46
section centroid (ybar) from XX=8.040816
2nd moment of area of section about its centroid=1864.114
```

EXAMPLES FOR PRACTICE 3

1. Determine the second moments of area about the centroid of the squares shown in Figs Q.3.1(a) and Q.3.1(b).

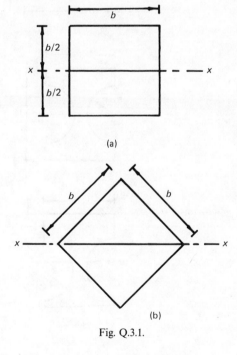

(a)

(b)

Fig. Q.3.1.

$\{b^4/12,\ b^4/12\}$.

2. Determine the positions of the centroidal axes x–x and the second moments of area about these axes, for the sections of Figs Q.3.2(a) to Q.3.2(d).

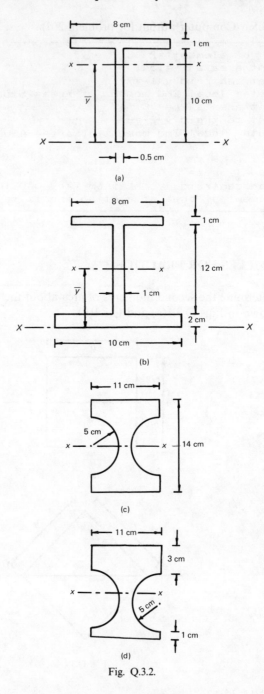

Fig. Q.3.2.

(a) {8.38 cm, 1.354×10^{-6} m^4}, (b) {5.8 cm, 1.275×10^{-5} m^4},
(c) {2.024×10^{-5} m^4}, (d) {8.04 cm, 1.864×10^{-5} m^4}.

3. Determine the second moment of area of the section shown in Fig. Q.3.3
about an axis passing through the centroid and parallel to the XX axis.

What would be the percentage reduction in second moment of area, if the bottom flange were identical to the top flange? (Portsmouth Polytechnic, June 1982)

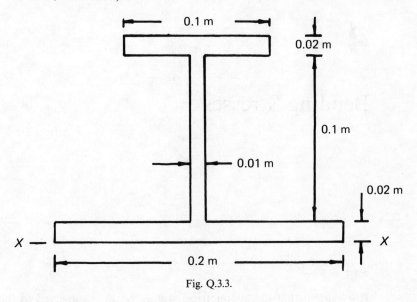

Fig. Q.3.3.

{2.058×10^{-5} m^4, 24.9%}.

4

Bending Stresses

4.1.1

It is evident that if a symmetrical section beam is subjected to the bending action shown in Fig. 4.1, the fibres at the top of the beam decrease their lengths, whilst the fibres at the bottom of the beam will increase their lengths. The effect of this will be to cause compressive direct stresses to occur in the fibres at the top of the beam and tensile direct stresses to occur in the fibres at the bottom of the beam, these stresses being parallel to the axis of the beam.

It is evident also from Fig. 4.1, that as the top layers of the beam decrease their lengths and the bottom fibres increase their lengths, that somewhere between the two, there will be a layer that will be in neither compression nor tension. This layer is called the *neutral layer*, and its intersection with the beam's cross-section is called the *neutral axis* (N.A.). The stress at the neutral axis is zero.

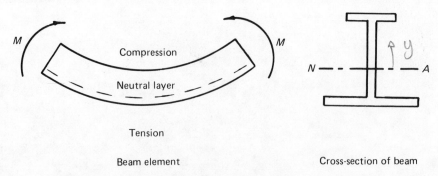

Beam element   Cross-section of beam

Fig. 4.1.   Beam element in bending.

128

4.1.2   ASSUMPTIONS MADE

The assumptions made in the theory of bending in this chapter are as follows:
(a) Deflections are small.
(b) The radius of curvature of the deformed beam is large compared to its other dimensions.
(c) The beam is initially straight.
(d) The cross-section of the beam is symmetrical.
(e) The effects of shear are negligible.
(f) Transverse sections of the beam, which are plane and normal before bending, remain plane and normal during bending.
(g) Elastic theory is obeyed, and the elastic modulus of the beam is the same in tension as it is in compression.
(h) The beam material is homogeneous and isotropic.

4.2.1

Proof of

$$\frac{\sigma}{y} = \frac{M}{I} = \frac{E}{R} \tag{4.1}$$

where,

σ = stress (due to the bending moment M) occurring at a distance y from the neutral axis.
M = bending moment.
I = second moment of area of the cross-section about its neutral axis (centroidal axis).
E = elastic modulus.
R = radius of curvature of the neutral layer of the beam, when M is applied.

Equation (4.1) is the fundamental expression that is used in the bending theory of beams, and it will be proven with the aid of Fig. 4.2, which shows the deformed shape of an initially straight beam under the action of a sagging bending moment M. Initially, all layers of the beam element will be the same length as the neutral layer, AB, so that,

Initial length of CD = AB = $R\theta$

Final length of CD = $(R + y)\theta$

where CD is the length of the beam element at a distance y from the neutral layer.

$$\text{Tensile strain of CD} = \frac{(R + y)\theta - R\theta}{R\theta}$$

$$= \frac{y}{R}$$

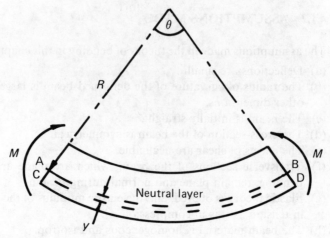

Fig. 4.2.   Beam element in bending.

and

stress in the layer CD = $\sigma = Ey/R$

Therefore

$$\frac{\sigma}{y} = \frac{E}{R} \tag{4.2}$$

Equation (4.2) shows that the bending stress σ varies linearly with y and it will act on the section, as shown in Fig. 4.3, where N.A. is the position of the neutral axis. Equation (4.2) also shows that the *largest stress* in magnitude occurs in the fibre which is the *furthest distance from the neutral axis*.

Maximum stress
is at fibre furthest
distance from
neutral
    axis

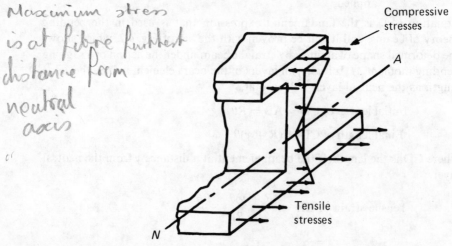

Fig. 4.3.   Bending stress distribution.

4.2.2

It is evident also, from equilibrium considerations, that the longitudinal tensile force caused by the tensile stresses, due to bending, must be equal and opposite to the longitudinal compressive force caused by the compressive stresses due to bending, so that,

$$\int \sigma \, dA = 0$$

where dA = area of element of cross-section at a distance y from the neutral axis but,

$$\sigma = Ey/R$$

therefore

$$\frac{E}{R} \int y \, dA = 0$$

i.e.

$$\int y \, dA = 0$$

i.e. *the neutral axis* is at the *centroid of the beam's cross-section.*

4.2.3

Now it can be seen from Fig. 4.3 that the bending stresses cause a couple which, from equilibrium considerations, must be equal and opposite to the externally applied moment M at the appropriate section, i.e.

$$M = \int \sigma y . dA$$

$$= \frac{E}{R} \int y^2 . dA$$

but,

$$\int y^2 . dA = I$$

therefore

$$\frac{M}{I} = \frac{E}{R} \tag{4.3}$$

From equations (4.2) and (4.3):

$$\frac{\sigma}{y} = \frac{M}{I} = \frac{E}{R} \tag{4.4}$$

4.2.4 SECTIONAL MODULUS (Z)

From equation (4.2), it can be seen that the maximum stress due to bending occurs in the fibre which is the greatest distance from the neutral axis.
  Let,

$\bar{y}$ = distance of the fibre in the cross-section of the beam which is the furthest distance from N.A.

$$Z = \frac{I}{\bar{y}} = \text{sectional modulus} \tag{4.5}$$

therefore maximum bending stress $= \hat{\sigma} = \dfrac{M}{Z}$ $\tag{4.6}$

4.3.1 EXAMPLE 4.1 BEAM UNDER PURE BENDING

A solid circular section steel bar of diameter 2 cm and length 1 m is subjected to a pure bending moment of magnitude M. If the maximum permissible stress in the bar is 100 MN/m², determine the maximum permissible value of M. If the lateral deflection at the mid-point of this beam, relative to its two ends, is 6.25 mm, what will be the elastic modulus of the beam?

$$I = \frac{\pi d^4}{64} = \frac{\pi \times (2 \times 10^{-2})^4}{64} = 7.854 \times 10^{-9} \text{ m}^4$$

The maximum stress will occur at the fibre in the cross-section which is the furthest distance from the neutral axis, *i.e.*

$$\bar{y} = 1 \text{ cm} = 1 \times 10^{-2} \text{ m}$$
$$Z = I/\bar{y} = 7.854 \times 10^{-7} \text{ m}^3$$

From equation (4.6):

$$M = \hat{\sigma}Z = 100 \times 10^6 \times 7.854 \times 10^{-7} = \underline{78.54 \text{ Nm}}$$

4.3.2

Now under pure bending, the beam will bend into a perfect arc of a circle, as shown in Fig. 4.4. In Fig. 4.4,

l = length of beam

δ = central deflection

Now from the properties of a circle,

$$\delta(2R - \delta) = \frac{l}{2} * \frac{l}{2}$$

or,

$$2R\delta - \delta^2 = l^2/4$$

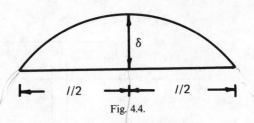

From properties of circle

Fig. 4.4.

but as deflections are small, δ^2 is small compared with $2R\delta$. Therefore

$\delta^2 \to 0$

$$\delta = l^2/(8R)$$

or,

$$R = l^2/(8\delta)$$

$$= 1/(8 \times 6.25 \times 10^{-3})$$

$$R = 20 \text{ m}$$

From equation (4.4):

$$E = MR/I = 78.54 \times 20/(7.854 \times 10^{-9})$$

$$E = 2 \times 10^{11} \text{ N/m}^2$$

4.4.1   EXAMPLE 4.2   BEAM WITH UDL

A beam of length 2 m and with the cross-section of Fig. 3.15 is simply-supported at its ends and carries a uniformly distributed load w, spread over its entire length, as shown in Fig. 4.5.

Determine a suitable value for w, given that the maximum permissible tensile stress is 100 MN/m² and the maximum permissible compressive stress is 30 MN/m².

The maximum bending moment $=$

$$\hat{M} = wl^2/8 \quad \text{at mid-span} \tag{4.7}$$

The bottom of the beam will be in tension, and the top will be in compression.

Now from equation (3.26), the distance of the neutral axis is 0.104 m from the bottom, as shown in Fig. 4.6. In Fig. 4.6,

y_1 is used to determine the maximum compressive stress, and
y_2 is used to determine the maximum tensile stress.

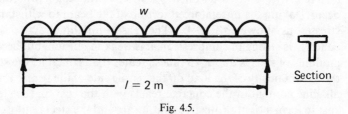

w

Section

$l = 2$ m

Fig. 4.5.

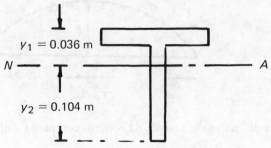

Fig. 4.6.   Beam cross-section.

4.4.2   To Determine the Design Criterion

If the tensile stress of 100 MN/m² is used in conjunction with y_2, then, the maximum permissible compressive stress, which is at the top

$$= \frac{0.036}{0.104} \times 100 = 34.6 \text{ MN/m}^2$$

i.e. the compressive stress in the top of the flange will be exceeded if the tensile stress of 100 MN/m² is adopted as the design criterion; hence, the design criterion is the 30 *MN/m² in the top flange*. If the top flange is under a compressive stress of 30 MN/m² then the tensile stress at the bottom of the web = 30 × 0.104/0.036 = 86.67 MN/m², in the bottom flange. Therefore

$$\hat{M} = \sigma * I/y = 30 \times 10^6 \times 5.02 \times 10^{-6}/0.036$$

$$\underline{\hat{M} = 4183.3 \text{ N m}} \tag{4.8}$$

Equating (4.7) and (4.8):

$$\underline{w = 8.37 \text{ kN/m}}$$

4.5.1   COMPOSITE BEAMS

Composite beams occur in a number of different branches of engineering, and appear in the form of reinforced concrete beams, flitched beams, ship structures, glass reinforced plastics, etc.

4.6.1

In the case of *reinforced concrete beams*, it is normal practice to reinforce the concrete with steel rods on the section of the beam where tensile stresses occur, leaving the unreinforced section of the beam to withstand compressive stresses, as shown in Fig. 4.7. The reason for this practice is that, whereas concrete is strong in compression, it is weak in tension, but because the elastic modulus of steel is about 15 times greater than concrete, the steel will absorb the vast majority of the load on the tensile side of the beam. Furthermore, the alkaline content in the concrete reacts with the rust on the steel, causing the rust to form a tight protective coating around the steel reinforcement, thereby

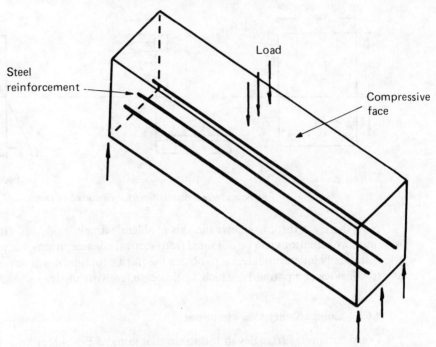

Fig. 4.7.   Reinforced concrete beam.

preventing further rusting. Thus, steel and concrete form a mutually compatible pair of materials which, when used together, actually improve each other's performance.

4.6.2

The method of analysing reinforced concrete beams of the type shown in Fig. 4.7 is to assume that all the tensile load is taken by the steel reinforcement and that all the compressive load is taken by that part of the concrete above the neutral axis, so that the stress distribution will be as shown in Fig. 4.8. In Fig. 4.8,

σ_s = tensile stress in steel

$\hat{\sigma}_c$ = maximum compressive stress in concrete

H = distance of the neutral axis of the beam from its top face

B = breadth of beam

D = distance between the steel reinforcement and the top of the beam

Let,

A_s = cross-sectional area of steel reinforcement

E_s = Young's modulus for steel

E_c = Young's modulus for concrete

$m = E_s/E_c$ = *modular ratio*

ε_s = tensile strain in steel

$\hat{\varepsilon}_c$ = maximum strain in concrete

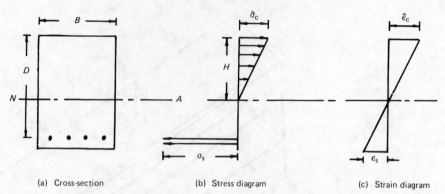

(a) Cross-section   (b) Stress diagram   (c) Strain diagram

Fig. 4.8.   Stress and strain distributions for reinforced concrete.

Now there are three unknowns in this problem, namely σ_s, $\hat{\sigma}_c$ and H, and as only two equations can be obtained from statical considerations, the problem is statically indeterminate, i.e. to obtain the third equation, it will be necessary to consider compatibility, which, in this case, consists of strains.

4.6.3   Compatibility Considerations

Consider similar triangles in the strain diagrams of Fig. 4.8(c).

$$\frac{\hat{\varepsilon}_c}{H} = \frac{\varepsilon_s}{(D - H)}$$

or,

$$\frac{\hat{\sigma}_c}{E_c H} = \frac{\sigma_s}{E_s(D - H)}$$

or,

$$\hat{\sigma}_c = \frac{E_c}{E_s} \frac{H\sigma_s}{(D - H)}$$

therefore

$$\hat{\sigma}_c = \frac{H\sigma_s}{m(D - H)} \tag{4.9}$$

4.6.4   Considering "Horizontal Equilibrium"

Tensile force in steel = Compressive force in concrete

$$\sigma_s A_s = \left(\frac{\hat{\sigma}_c}{2}\right) BH$$

or,

$$\hat{\sigma}_c = \frac{2\sigma_s A_s}{BH} \tag{4.10}$$

4.6.5   Considering "Rotational Equilibrium"

Externally applied moment M = Moment of resistance of the section or,

$$M = \sigma_s A_s * (D - H) + \left(\frac{\hat{\sigma}_c}{2}\right) * BH * \frac{2H}{3}$$

$$\underline{M = \sigma_s A_s(D - H) + \hat{\sigma}_c BH^2/3} \tag{4.11}$$

Equation (4.9) and (4.10):

$$\frac{H\sigma_s}{m(D - H)} = \frac{2\sigma_s A_s}{BH}$$

or,

$$BH^2 = 2m(D - H)A_s$$

therefore

$$H^2 + 2m A_s H/B - 2m D A_s/B = 0$$

i.e.

$$H = \frac{-2m A_s/B + \sqrt{\{(2m A_s/B)^2 + 8m D A_s/B\}}}{2}$$

$$\underline{H = \sqrt{\{(m A_s/B)^2 + 2m D A_s/B\}} - m A_s/B} \tag{4.12}$$

Substituting equation (4.10) into equation (4.11):

$$M = \sigma_s A_s(D - H) + \frac{2\sigma_s A_s}{BH}\frac{BH^2}{3}$$

therefore

$$\sigma_s = \frac{M}{A_s\{(D - H) + 2H/3\}} = \frac{M}{A_s(D - H/3)} \tag{4.13}$$

and from equation (4.10):

$$\hat{\sigma}_c = \frac{2M}{BH(D - H/3)} \tag{4.14}$$

4.7.1   EXAMPLE 4.3   REINFORCED CONCRETE BEAM

A reinforced concrete beam, of rectangular section, is subjected to a bending moment, such that the steel reinforcement is in tension.

Given the following, determine the maximum permissible value of this bending moment.

$$D = 0.4\,\text{m} \qquad\qquad B = 0.3\,\text{m} \qquad\qquad m = \;\; 15$$

Maximum permissible compressive stress in concrete = $10\ \text{MN/m}^2$
Maximum permissible tensile stress in steel   = $150\ \text{MN/m}^2$
Diameter of each steel reinforcing rod   = $2\ \text{cm}$

n = no. of rods = 8

$$A_s = 2.513 \times 10^{-3} \text{ m}^2$$

From equation (4.12):

$$H = \sqrt{\{0.0158 + 0.1005\}} - 0.12565$$

$$\underline{H = 0.215 \text{ m}}$$

From equation (4.13):

$$M = \sigma_s A_s (D - H/3)$$

$$\underline{M = 0.124 \text{ MN m}}$$

From equation (4.14):

$$M = \hat{\sigma}_c BH(D - H/3)/2$$

$$\underline{M = 0.106 \text{ MN m}}$$

i.e. the maximum stress in the concrete is the design criterion. Therefore

Maximum permissible bending moment = 0.106 MN m

N.B. The overall dimensions of the beam's cross-section should allow for the steel reinforcement to be covered by at least 5 cm of concrete.

4.8.1 FLITCHED BEAMS

A flitched beam is a common type of composite beam, where the reinforcements are relatively thin compared to the depth of the beam, and are usually attached to its outer surfaces, as shown in Fig. 4.9. Typical materials used for flitched beams include a wooden core combined with external steel reinforcement and various types of plastic reinforcement combined with a synthetic porous core of low density.

Let,

M = applied moment at the section
M_r = moment of resistance of external reinforcement
M_c = moment of resistance of core,

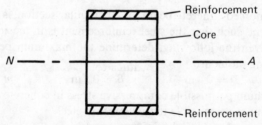

Fig. 4.9.   Flitched beam.

so that,

$$M = M_r + M_c \qquad (4.15)$$

The main assumption made is that the *radius of curvature, R, is the same for the core as it is for the reinforcement*, i.e.

$$R = \frac{E_c I_c}{M_c} = \frac{E_r I_r}{M_r}$$

or,

$$M_r = \frac{E_r I_r}{E_c I_c} M_c \qquad (4.16)$$

where,

I_r = second moment of area of the external reinforcement about the neutral axis of the composite beam.

I_c = second moment of area of the core about the neutral axis of the composite beam.

E_r = Young's modulus for the external reinforcement.

E_c = Young's modulus for the core.

Substituting equation (4.16) into (4.15):

$$M = \frac{M_c}{E_c I_c} (E_r I_r + E_c I_c)$$

therefore

$$M_c = \frac{E_c I_c M}{(E_r I_r + E_c I_c)} \qquad (4.17)$$

and,

$$M_r = \frac{E_r I_r M}{(E_r I_r + E_c I_c)} \qquad (4.18)$$

Now,

$$\sigma_r = M_r / Z_r$$

and,

$$\sigma_c = M_c / Z_c$$

where,

σ_r = maximum stress in the external reinforcement

σ_c = maximum stress in the core

Z_r = sectional modulus of the external reinforcement about N.A.

Z_c = sectional modulus of the core

Hence,

$$\sigma_r = E_r y_r M / (E_r I_r + E_c I_c) \qquad (4.19)$$

and,

$$\sigma_c = E_c \, y_c \, M/(E_r \, I_r + E_c \, I_c) \qquad (4.20)$$

where,

y_r = distance of the outermost fibre of the external reinforcement from the neutral axis of the composite beam.

y_c = distance of the outermost fibre of the core from the neutral axis of the composite beam.

4.9.1   EXAMPLE 4.4   FLITCHED BEAM

(a) A wooden beam of rectangular section is of depth 10 cm and width 5 cm. Determine the moment of resistance of this section given the following:

Young's modulus for wood = 1.4×10^{10} N/m²
Maximum permissible stress in wood = 20 MN/m²

(b) What percentage increase will there be in the moment of resistance of the beam section if it is reinforced by two 5 mm thick galvanised steel plates attached to the top and bottom surfaces of the beam?

Young's modulus for steel = 2×10^{11} N/m²
Maximum permissible stress in steel = 150 MN/m²

4.9.2

(a) Now,

$$M = \frac{\sigma I}{y}$$

I_w = second moment of area of wood about its neutral axis
= $5 \times 10^{-2} \times (10 \times 10^{-2})^3/12$

$I_w = 4.167 \times 10^{-6}$ m⁴

$\bar{y} = 5 \times 10^{-2}$ m

therefore

M_w = moment of resistance of wood

$$= \frac{20 \times 10^6 \times 4.167 \times 10^{-6}}{5 \times 10^{-2}}$$

$\underline{M_w = 1666.7 \text{ N m}}$

4.9.3

(b) R_w = radius of curvature of wood

R_s = radius of curvature of steel

but,

$$R_s = R_w$$

therefore,

$$\frac{\sigma_s}{E_s\, y_s} = \frac{\sigma_w}{E_w\, y_w}$$

or,

$$\sigma_s = \sigma_w \cdot \frac{E_s \cdot y_s}{E_w \cdot y_w}$$

$$= \frac{\sigma_w \times 2 \times 10^{11} \times 5.25 \times 10^{-2}}{1.4 \times 10^{10} \times 5 \times 10^{-2}}$$

$$\underline{\sigma_s = 15\, \sigma_w}$$

i.e.

$$\underline{\sigma_s = 150 \text{ MN/m}^2 \text{ is the design criterion}}$$

where,

σ_s = maximum stress in steel
E_s = elastic modulus of steel
E_w = elastic modulus of wood
y_s = distance of steel from N.A.
y_w = distance of outermost fibre of wooden core from N.A.

Now, as

$$\sigma_s = 15\, \sigma_w$$

$$\sigma_w = \tfrac{150}{15} = 10 \text{ MN/m}^2$$

therefore

$$M_w = 1666.7 \times 10/20 = \underline{833.4 \text{ N m}}$$

$$I_s = 5 \times 10^{-2} \times 5 \times 10^{-3} \times (5.25 \times 10^{-2})^2 \times 2$$
$$= 1.378 \times 10^{-6}$$

and,

$$M_s = \frac{150 \times 10^6 \times 1.378 \times 10^{-6}}{5.25 \times 10^{-2}} = \underline{3937.5 \text{ N m}}$$

$$M = M_s + M_w = \underline{4770.9 \text{ N m}}$$

i.e. percentage increase in moment of resistance of the flitched beam over the wooden beam = 186.2.

N.B.   In this case, the chosen thickness for the steel plate was too small, as the stress in the wood was well below its permissible value.

4.10.1   COMPOSITE SHIP STRUCTURES

Composite ship structures appear in the form of a steel hull, together with an aluminium alloy superstructure. The reason for this combination is that steel is a suitable material for the main hull of a ship because of its ductility and good welding properties, but in order to keep the centre of gravity of a ship as low as possible, for the purposes of ship stability, it is convenient to use a material with a lower density than steel for the superstructure. In general, it is not suitable to use aluminium for the main hull, because aluminium has poor corrosion resistance to salt water.

Under longitudinal bending moments, caused by the self-weight of the ship and buoyant forces, due to waves, as shown in Fig. 4.10, the longitudinal strength of a ship can be based on beam theory, where the cross-section of the

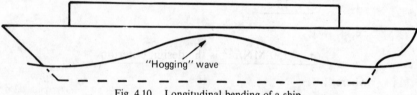

Fig. 4.10.   Longitudinal bending of a ship.

"equivalent beam" is in fact the cross-section of the ship, as shown in Fig. 4.11. The strain distribution across the transverse section of the ship is as shown in Fig. 4.11(b), but as stress $= E *$ strain, the equivalent moment of resistance of the aluminium alloy will be equivalent to E_a/E_s of a steel section of the same size. Thus, to calculate the position of the neutral axis (N.A.), and the second moment of area, the aluminium alloy superstructure can be assumed to be equivalent to the form shown in Fig. 4.12, where,

E_a = Young's modulus for aluminium alloy
E_s = Young's modulus for steel

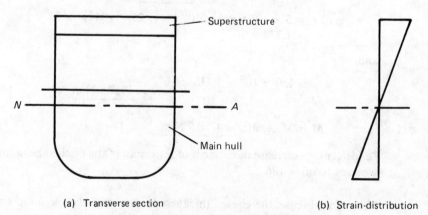

(a)   Transverse section   (b)   Strain-distribution

Fig. 4.11. Cross-section of ship.

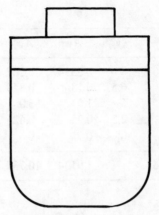

Fig. 4.12.   Equivalent section.

4.10.2 ¯EXAMPLE 4.5   COMPOSITE "SHIP" TYPE STRUCTURE

A box-like cross-section consists of two parts, namely a steel bottom and an aluminium alloy top, as shown in Fig. 4.13. If the plate thickness of both the aluminium alloy and the steel is 1 cm, determine the maximum stress in both materials, when the section is subjected to a bending moment of 100 MN m, which causes it to bend about a horizontal plane (N.A.).

E_a = Young's modulus for aluminium alloy = 6.67×10^{10} N/m²
E_s = Young's modulus for steel   = 2×10^{11} N/m²

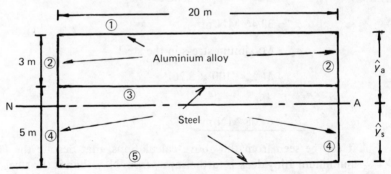

Fig. 4.13.   Cross-section of composite structure.

4.10.3

To determine the position of the neutral axis and the second moment of area of the equivalent steel section, use will be made of Table 4.1.

$$\hat{y}_s = \sum Ay / \sum A = 1.9134/0.5867$$
$$\underline{\hat{y}_s = 3.261 \text{ m}}$$

Table 4.1

Section	A	y	Ay	Ay^2	i
①	0.0667	8	0.5334	4.267	$20 \times (1 \times 10^{-2})^3/36 = 0$
②	0.02	6.5	0.13	0.845	$1 \times 10^{-2} \times 3^3 \times 2/36 = 0.015$
③	0.2	5	1.0	5.0	$20 \times (1 \times 10^{-2})^3/12 = 0$
④	0.1	2.5	0.25	0.625	$1 \times 10^{-2} \times 5^3 \times 2/12 = 0.208$
⑤	0.2	0	0	0	$(1 \times 10^{-2})^3 \times 20/3 = 0$
$\sum$	0.5867	—	1.9134	10.737	0.223

therefore

$$\hat{y}_a = 4.739 \text{ m}$$

$$I_{XX} = \sum Ay^2 + \sum i = 10.737 + 0.223$$

$$\underline{I_{XX} = 10.96 \text{ m}^4}$$

$$I_{NA} = I_{XX} - (\hat{y}_s)^2 \sum A = 10.96 - 3.261^2 \times 0.5867$$

$$\underline{I_{NA} = 4.721 \text{ m}^4}$$

$\hat{\sigma}_a$ = Maximum stress in the aluminium alloy

$$= \frac{M\,\hat{y}_a}{I_{NA}} * \left(\frac{E_a}{E_s}\right) = \frac{100 \times 4.739}{4.721 \times 3}$$

$$\hat{\sigma}_a = 33.46 \text{ MN/m}^2$$

$\hat{\sigma}_s$ = Maximum stress in the steel

$$= \frac{M\,\hat{y}_s}{I_{NA}} = \frac{100 \times 3.261}{4.721}$$

$$\hat{\sigma}_s = 69.08 \text{ MN/m}^2$$

It can be seen from the above calculations, that despite the fact that the aluminium alloy deck is further away from N.A. than is the steel bottom, the stress in the aluminium alloy is less than the steel, because its elasticity is three times greater than the steel.

4.11.1 Glass Reinforced Plastic

Glass reinforced plastics are of much interest in structures varying from car bodies to boat hulls, and from chairs to ship superstructures. Analysis of structures composed of glass reinforced plastics is beyond the scope of this book, and for further study the reader should consult reference [2].

4.12.1   Conclusions on Composite Structures

This chapter has shown that the sensible use of composites can improve the structural efficiency of many types of structure in various engineering applications.

4.13.1   COMBINED BENDING AND DIRECT STRESS

The case of combined bending and direct stress occurs in a number of engineering situations, including the eccentric loading of short columns, as shown in Fig. 4.14. By placing the equal and opposite forces on the centre line of the strut, as shown in Fig. 4.14(b), the loading condition of Fig. 4.14(a) is unaltered. However, it can be seen that the column of Fig. 4.14(b) is in fact subjected to a centrally applied force P and a couple $P\Delta$, as shown in Fig. 4.14(c). Furthermore, from Fig. 4.14(c), it can be seen that due to the centrally applied direct load P, the whole of the strut will be subjected to a direct compressive σ_d, but due to the couple $P\Delta$, the side AB will be in tension and the side CD will be in compression.

Thus, the effect of M will be to cause the stress to be increased in magnitude on face CD, and to be decreased in magnitude on face AB, i.e.

$$\text{Stress on face AB} = \sigma_{AB} = -\frac{P}{A} + \frac{M\bar{y}}{I} \qquad (4.21)$$

$$\text{Stress on face CD} = \sigma_{CD} = -\frac{P}{A} - \frac{M\bar{y}}{I} \qquad (4.22)$$

It is evident, therefore, that in general, the stress due to the combined effects of a bending moment M and a tensile load P will be given by

$$\sigma = \sigma_d \pm \sigma_b$$

where,

σ_d = direct stress (tensile is positive)
σ_b = bending stress

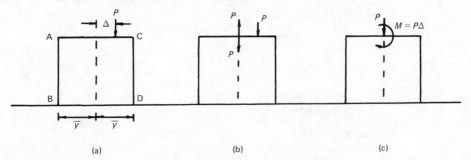

(a)   (b)   (c)

Fig. 4.14.   Eccentrically loaded short column.

4.13.2 ECCENTRICALLY LOADED CONCRETE COLUMNS

As concrete is weak in tension, it is desirable to determine how eccentric a load can be so that no part of a short column is in tension.

The following two examples will be used to demonstrate the calculations usually associated with eccentrically loaded short columns.

4.14.1 EXAMPLE 4.6 ECCENTRICALLY LOADED SQUARE SECTION COLUMN

Determine the position in which an eccentrically applied vertical compressive load can be placed, so that no tension occurs in a short vertical column of square cross-section.

By applying a compressive force P at the point shown in Fig. 4.15, it can be seen that the face AB is likely to develop tensile stresses due to the bending action about the YY axis.

MID-THIRD RULE

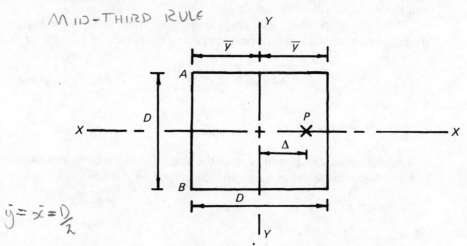

$\bar{y} = \bar{x} = D/2$

Fig. 4.15. Cross-section of concrete column.

To satisfy the requirements of the example, let the stress on the face AB equal zero, so that,

$$0 = -\frac{P}{A} + \frac{P\Delta \bar{y}}{I}$$

$$= -\frac{1}{D^2} + \frac{6\Delta}{D^3}$$

Therefore

$$\Delta = D/6 \tag{4.23}$$

For no tension must have 0 stress

From equation (4.23), it can be seen that for no tension to occur in the short column, owing to an eccentrically applied compressive load, the eccentrically applied load must be applied within the mid-third area of the centre of the square section. For this reason, this rule is known as the *mid-third rule*.

4.14.2   EXAMPLE 4.7   ECCENTRICALLY LOADED CIRCULAR SECTION COLUMN

Determine the position in which an eccentrically applied vertical compressive load can be placed, so that no tension occurs in a short vertical column of circular cross-section.

MID-QUARTER RULE

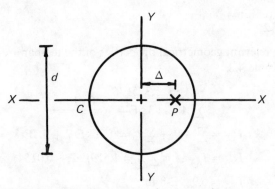

Fig. 4.16.   Cross-section of concrete column.

By applying a compressive force P to the point shown in Fig. 4.16, it can be seen that tension is most likely to occur at the point "C", so that to satisfy the requirements of the example, the stress at "C" equals zero.

$$0 = -\frac{4P}{\pi d^2} + 32\,\frac{P\Delta}{\pi d^3}$$

or,

$$\Delta = d/8 \qquad\qquad (4.24)$$

Equation (4.24) shows that for no tensile stress to occur in a short concrete column of circular cross-section, the eccentricity of the load must not exceed $d/8$. For this reason, this is known as the *mid-quarter rule*.

4.15.1   EXAMPLE 4.8   CLAMP UNDER LOADING

Determine the maximum tensile and compressive stresses in the clamp of Fig. 4.17.

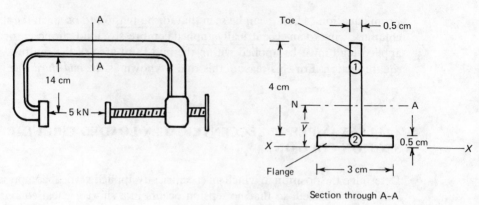

Fig. 4.16.   Clamp under loading.

4.15.2

The relevant geometrical properties of the tee-bar are calculated with the aid of Table 4.2.

$$\bar{y} = \frac{\sum Ay}{\sum A} = \frac{5.375\text{E-}6}{3.5\text{E-}4} = \underline{0.0154 \text{ m}}$$

$$I_{XX} = \sum Ay^2 + \sum i_0 = 1.259\text{E-}7 + 2.7\text{E-}8 = \underline{1.529\text{E-}7 \text{ m}^4}$$

$$I_{NA} = I_{XX} - \bar{y}^2 \sum A = 1.529\text{E-}7 - 0.0154^2 \times 3.5\text{E-}4$$

$$\underline{I_{NA} = 6.989\text{E-}8 \text{ m}^4}$$

Table 4.2.   Geometrical properties of the tee-bar.

Section	$A(\text{m}^2)$	$y(\text{m})$	$Ay(\text{m}^3)$	$Ay^2(\text{m}^4)$	$i_0(\text{m}^4)$
①	2E-4	2.5 E-2	5E-6	1.25 E-7	2.667E-8
②	1.5E-4	0.25E-2	3.75 E-7	9.38 E-10	3.125E-10
$\sum$	3.5E-4	—	5.375E-6	1.259E-7	2.7 E-8

4.15.3

Let,

$\hat{M}$ = the maximum bending moment on the "top" member of clamp (or at A–A)

$$= 5 \text{ kN} \times (14 \times 10^{-2} \text{ m} + \bar{y})$$

$$= 5 \text{ kN} \times (14 \times 10^{-2} + 0.0154)$$

$$\underline{\hat{M} = 0.777 \text{ kN m}}$$

σ_T = the stress in the "top" (toe) of the tee

$$= -\frac{0.777 \text{ kN m} \times 0.0296 \text{ m}}{6.989\text{E-8 m}^4} + \frac{5 \text{ kN}}{A}$$

$$= -329077 \frac{\text{kN}}{\text{m}^2} + \frac{5 \text{ kN}}{3.5\text{E-4 m}^2}$$

$$\sigma_T = -329077 + 14286 = -3147491 \text{ kN/m}^2$$

$$\underline{\sigma_T = -314.8 \text{ MN/m}^2}$$

σ_B = the maximum stress in the "bottom" (flange) of the tee

$$= \frac{0.777 \times 0.0154}{6.989\text{E-8}} + 14286 = 171209 + 14286$$

$$= 185495 \text{ kN/m}^2$$

$$\underline{\sigma_B = 185.5 \text{ MN/m}^2}$$

N.B.   By placing the flange of the tee-bar on the inner part of the clamp, the maximum stresses have been reduced. This has been achieved by "lowering" the neutral axis towards the tensile face, where the bending and direct stresses are additive. Movement of the neutral axis towards the tensile face also had the effect of lowering the maximum bending moment, as the lever arm that the load was acting on was decreased.

EXAMPLES FOR PRACTICE 4

1. A concrete beam of uniform square cross-section, as shown in Fig. Q.4.1, is to be lifted by its ends, so that it may be regarded as being equivalent to a horizontal beam, simply-supported at its ends and subjected to a uniformly distributed load, due to its self-weight.

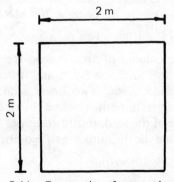

Fig. Q.4.1.   Cross-section of concrete beam.

Determine the maximum permissible length of this beam, given the following:

Density of concrete = 2400 kg/m³

Maximum permissible tensile stress in the concrete = 1 MN/m²

$$g = 9.81 \text{ m/s}^2$$

{10.64 m}

2. If the concrete beam of Example 1 had a hole in the bottom of its cross-section, as shown in Fig. Q.4.2, what would be the maximum permissible length of the beam?

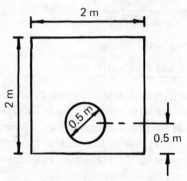

Fig. Q.4.2.   Cross-section with hole.

{10.55 m}

3. What would be the maximum permissible length of the beam, if the hole were at the top?
   {10.83 m}

4. A horizontal beam, of length 4 m, is simply-supported at its ends and subjected to a vertically applied concentrated load of 10 kN at mid-span. Assuming that the width of the beam is constant and equal to 0.03 m, and neglecting the self-weight of the beam, determine an equation for the depth of the beam, so that the beam will be of uniform strength. The maximum permissible stress in the beam = 100 MN/m².
   {$d = 0.1 x^{1/2}$ from 0 to 2 m}

5. A short steel column, of circular cross-section, has an external diameter of 0.4 m and wall thickness 0.1 m and it carries a compressive but axially applied eccentric load. Two linear strain gauges, which are mounted longitudinally at opposite sides on the external surface of the column, but in the plane of the load, record strains of 50×10^{-6} and -200×10^{-6}.
   Determine the magnitude and eccentricity of the axial load.

   $$E = 2 \times 10^{11} \text{ N/m}^2$$

   {−1.413 MN; 0.104 m}

6. Determine the maximum tensile and compressive stresses in the clamp of
   Fig. Q.4.6.

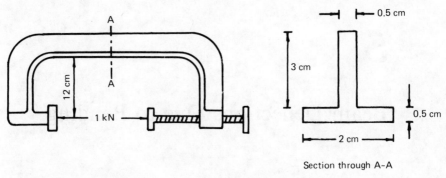

Fig. Q.4.6.

$$\{61.9 \text{ MN/m}^2; \ -94 \text{ MN/m}^2\}$$

7. The cross-section of a reinforced concrete beam is as shown in Fig. Q.4.7.
   Determine the maximum bending moment that this beam can sustain,
   assuming that the steel reinforcement is on the tensile side and that the
   following apply:

Maximum permissible compressive stress in concrete	$= 10 \text{ MN/m}^2$
Maximum permissible tensile stress in steel	$= 200 \text{ MN/m}^2$
Modular ratio	$= 15$
Diameter of a steel rod	$= 2 \text{ cm}$
n = no. of steel rods	$= 6$

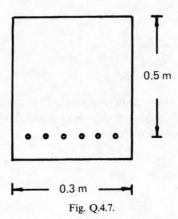

Fig. Q.4.7.

$$\{0.144 \text{ MN m}\}$$

5

Beam Deflections Due to Bending

5.1.1 DERIVATION OF EQUATION RELATING DEFLECTION AND BENDING MOMENT

Beam deflections are usually due to bending and shear, but only those due to the former will be considered in this chapter.

The radius of curvature R of a beam in terms of its deflection y, at a distance x along the length of the beam, is given by:

$$\frac{1}{R} = \frac{\dfrac{d^2 y}{dx^2}}{\left\{ 1 + \left(\dfrac{dy}{dx} \right)^2 \right\}^{3/2}} \tag{5.1}$$

However, if the deflections are small, as is the usual requirement in structural design, then $(dy/dx)^2$ is negligible compared with unity, so that equation (5.1) can be approximated by

$$\frac{1}{R} = \frac{d^2 y}{dx^2} \tag{5.2}$$

but,

$$\frac{M}{I} = \frac{E}{R}$$

therefore

$$EI \frac{d^2 y}{dx^2} = M \tag{5.3}$$

152

Equation (5.3) is a very important expression for the bending of beams, and there are a number of different ways of solving it, but only three methods will be considered here, namely by repeated integation, by moment-area methods and by use of the slope–deflection equations.

5.2.1   Repeated Integration Method

This is a boundary value method, which depends on integrating equation (5.3) twice and then substituting boundary conditions to determine the arbitrary constants, together with other unknowns. It will be demonstrated through detailed solutions of the following selection of examples.

5.3.1   EXAMPLE 5.1   CANTILEVER WITH END LOAD

Determine an expression for the maximum deflection of a cantilever of uniform section, under a concentrated load at its free end, as shown in Fig. 5.1.

$$EI\,\frac{\mathrm{d}^2 y}{\mathrm{d}x^2} = M$$

$$= -Wx$$

$$EI\,\frac{\mathrm{d}y}{\mathrm{d}x} = -\frac{Wx^2}{2} + A \tag{5.4}$$

$$EIy = -\frac{Wx^3}{6} + Ax + B \tag{5.5}$$

There are two unknowns, namely the arbitrary constants A and B; therefore, two boundary conditions will be required, as follows:

At $x = l, \dfrac{\mathrm{d}y}{\mathrm{d}x} = 0$ (i.e. the slope at the built-in end is zero).

Hence, from equation (5.4):

$$0 = -\frac{Wl^2}{2} + A$$

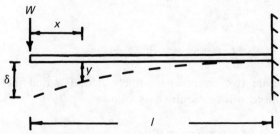

Fig. 5.1.   Cantilever with end load.

therefore

$$A = Wl^2/2 \tag{5.6}$$

The other boundary condition is that,

At $x = l, y = 0$ (i.e. the deflection y is zero at the built-in end).

Hence from equation (5.5):

$$0 = -\frac{Wl^3}{6} + Al + B$$

or

$$\underline{B = -Wl^3/3} \tag{5.7}$$

Substituting equations (5.6) and (5.7) into equation (5.5), the deflection y at any distance x along the length of the beam is given by

$$\underline{y = -\frac{W}{EI}(x^3/6 - l^2 x/2 + l^3/3)} \tag{5.8}$$

By inspection, the maximum deflection δ occurs at $x = 0$, i.e.

$$\delta = -\frac{Wl^3}{3EI} \tag{5.9}$$

The negative sign denote that the deflection is downward.

5.4.1　EXAMPLE 5.2　CANTILEVER WITH UDL

Determine an expression for the maximum deflection of a cantilever under a uniformly distributed load w, as shown in Fig. 5.2.

At any distance x along the length of the beam, the bending moment M is given by

$$M = -wx^2/2,$$

so that,

$$EI\frac{d^2 y}{dx^2} = -\frac{wx^2}{2} \tag{5.10}$$

$$EI\frac{dy}{dx} = -\frac{wx^3}{6} + A$$

$$EIy = -\frac{wx^4}{24} + Ax + B \tag{5.11}$$

There are two unknowns, namely A and B and therefore two boundary conditions will be required, as follows:

At $x = l, \dfrac{dy}{dx} = 0$

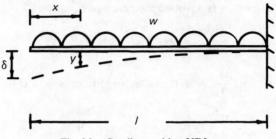

Fig. 5.2.   Cantilever with a UDL.

therefore

$$A = wl^3/6 \tag{5.12}$$

At $x = l$, $y = 0$

or,

$$0 = -\frac{wl^4}{24} + \frac{wl^4}{6} + B$$

therefore

$$B = -wl^4/8 \tag{5.13}$$

Substituting equations (5.12) and (5.13) into equation (5.11), the following expression is obtained for the deflection y of the beam at any distance x from its free end:

$$y = -\frac{w}{EI}\,(x^4/24 - l^3x/6 + l^4/8) \tag{5.14}$$

By inspection, the maximum deflection δ occurs at $x = 0$, where,

$$\delta = -wl^4/(8EI) \tag{5.15}$$

The negative sign denotes that the deflection is downward.

5.4.2   Alternative Method for Determining δ

From equation (1.46):

$$\frac{\mathrm{d}^2M}{\mathrm{d}x^2} = w$$

but,

$$EI\,\frac{\mathrm{d}^2y}{\mathrm{d}x^2} = M$$

therefore

$$EI\,\frac{\mathrm{d}^4y}{\mathrm{d}x^4} = w \tag{5.16}$$

In this case, *w is downward*; therefore equation (5.16) becomes

$$EI\,\frac{d^4y}{dx^4} = -w$$

$$EI\,\frac{d^3y}{dx^3} = F \text{ (the shearing force)} = -wx + A$$

$$EI\,\frac{d^2y}{dx^2} = M = -\frac{wx^2}{2} + Ax + B$$

At $x = 0, F = 0$;   therefore $A = 0$

At $x = 0, M = 0$;   therefore $B = 0$

i.e.

$$EI\,\frac{d^2y}{dx^2} = -\frac{wx^2}{2} \tag{5.17}$$

which is identical to equation (5.10).

Equation (5.16) is particularly useful for beams with distributed loads of complex form, such as those met in determining the longitudinal strengths of ships, owing to the combined effects of self-weight and buoyant forces caused by waves.

5.5.1   EXAMPLE 5.3   SIMPLY-SUPPORTED BEAM WITH CENTRALLY PLACED CONCENTRATED LOAD

Determine an expression for the maximum deflection of a uniform section beam, simply-supported at its ends and subjected to a centrally placed concentrated load *W*, as shown in Fig. 5.3.

In this case, there is a discontinuity at mid-span; hence, equation (5.3), together with its boundary conditions, can only be applied between $x = 0$ and $x = l/2$, i.e.

$$EI\,\frac{d^2y}{dx^2} = M = \frac{W}{2}\,x$$

$$EI\,\frac{dy}{dx} = \frac{Wx^2}{4} + A$$

$$EIy = \frac{Wx^3}{12} + Ax + B \tag{5.18}$$

At $x = 0, y = 0$;

therefore

$$0 = B$$

At $x = l/2, \dfrac{dy}{dx} = 0$

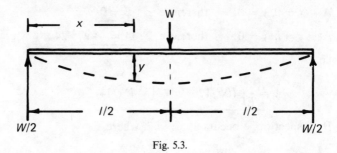

Fig. 5.3.

therefore

$$A = - Wl^2/16$$

i.e.

$$y = \frac{W}{EI} \ (x^3/12 - l^2 x/16)$$

By inspection, the maximum deflection δ occurs at $x = l/2$, where,

$$\delta = \frac{- Wl^3}{48EI} \tag{5.19}$$

5.6.1   EXAMPLE 5.4   SIMPLY-SUPPORTED BEAM WITH UDL

Determine an expression for the maximum deflection of a uniform section beam, simply-supported at its ends and subjected to a uniformly distributed load w, as shown in Fig. 5.4.

At $\quad x, M = \dfrac{wlx}{2} - \dfrac{wx^2}{2}$,

so that,

$$EI \frac{d^2 y}{dx^2} = \frac{wl}{2} x - \frac{wx^2}{2} \tag{5.20}$$

$$EI \frac{dy}{dx} = \frac{wlx^2}{4} - \frac{wx^3}{6} + A$$

$$EIy = \frac{wlx^3}{12} - \frac{wx^4}{24} + Ax + B$$

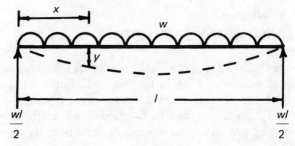

Fig. 5.4.

At $x = 0$, $y = 0$;   therefore $\underline{B = 0}$

At $x = l$, $y = 0$;   therefore $\underline{A = -wl^3/24}$

i.e.

$$y = \frac{w}{EI}\,(lx^3/12 - x^4/24 - l^3x/24) \tag{5.21}$$

By inspection, δ occurs at $x = l/2$, where,

$$\delta = -\frac{5wl^4}{384EI}$$

5.6.2   Alternative Method of Solving Example 5.4

$$EI\,\frac{d^4y}{dx^4} = -w$$

$$EI\,\frac{d^3y}{dx^3} = F = -wx + A$$

At $x = 0$, $F = wl/2$

therefore

$$\underline{A = wl/2}$$

or,

$$EI\,\frac{d^3y}{dx^3} = -wx + \frac{wl}{2}$$

$$EI\,\frac{d^2y}{dx^2} = M = -\frac{wx^2}{2} + \frac{wlx}{2} + B$$

At $x = 0$, $M = 0$;   therefore $\underline{B = 0}$

Therefore

$$EI\,\frac{d^2y}{dx^2} = -\frac{wx^2}{2} + \frac{wlx}{2} \tag{5.22}$$

which is identical to equation (5.20).

5.7.1   MACAULAY'S METHOD

This method will be given without proof, as a number of proofs already exist in numerous texts [3, 4]. In the case of Example 5.3, it can be seen that the equation for M only applied between $x = 0$ and $x = l/2$, and that both boundary conditions had to be applied within these limits. Furthermore, because of symmetry, the boundary condition, $dy/dx = 0$ at $x = l/2$, also applied.

However, if the beam were not symmetrically loaded, it would not be possible to obtain the second boundary condition, namely $dy/dx = 0$ at $x = l/2$.

To demonstrate how to overcome this problem, the following examples will be considered, which are based on Macaulay's method.

5.8.1 EXAMPLE 5.5 SIMPLY-SUPPORTED BEAM

Determine an expression for the deflection under the load for the uniform section beam of Fig. 5.5.

First, it will be necessary to determine the value of R_A, which can be obtained by taking moments about the point "B".

$$R_A l = Wb$$

therefore

$$\underline{R_A = Wb/l}$$

The method of Macaulay is to use separate bending moment equations for each section of the beam, but to integrate the equations via the Macaulay brackets, so that the constants of integration apply to all sections of the beam. It must, however, be emphasised that if the term within the *Macaulay bracket is negative*, then *that part of the expression does not apply* for boundary conditions, etc.

For the present problem, the bending moment between the points A and C will be different to that between the points C and B; hence, it will be necessary to separate the two expressions by the dashed line, as shown below.

$$EI \frac{d^2y}{dx^2} = R_A x \qquad \Big| \quad -W[x-a]$$

$$EI \frac{d^2y}{dx^2} = \frac{Wbx}{l} \qquad \Big| \quad -W[x-a] \qquad (5.23)$$

$$EI \frac{dy}{dx} = \frac{Wbx^2}{2l} + A \qquad \Big| \quad -\frac{W}{2}[x-a]^2 \qquad (5.24)$$

$$EIy = \frac{Wbx^3}{6l} + Ax + B \quad \Big| \quad -\frac{W}{6}[x-a]^3 \qquad (5.25)$$

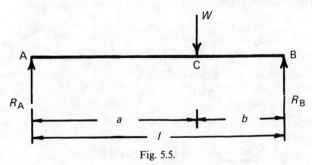

Fig. 5.5.

The brackets, [ ], which appear in equations (5.23) to (5.25) are known as Macauley brackets, and their integration must be carried out in the manner shown in equations (5.24) and (5.25), so that the arbitrary constants A and B apply to both sides.

Now in setting the boundary conditions and in obtaining values for dy/dx and y, *if the terms within the Macaulay brackets become negative, then they do not apply.*

5.8.2 Boundary Conditions

The first boundary condition is as follows:

At $x = 0$, $y = 0$ which, when applied to equation (5.25), reveals that $\underline{B = 0}$.

N.B. The expression $[0 - a]^3$ does not apply when the above boundary condition is substituted into equation (5.25), because the term within the Macaulay brackets, [ ], is negative.

The second boundary condition is as follows:

At $x = l$, $y = 0$

therefore

$$0 = \frac{Wbl^2}{6} + Al - \frac{Wb^3}{6}$$

or,

$$\underline{A = -\frac{Wab(l + b)}{l}}$$

i.e. the deflection y at a distance x along the length of the beam is given by equation (5.26), providing the term within the Macaulay brackets, [ ], does not become negative.

$$EIy = \frac{Wbx^3}{6l} - \frac{Wab(l + b)x}{l} \ \bigg| \ -\frac{W}{6}[x - a]^3 \qquad (5.26)$$

The deflection under the load δ_c is given by

$$\delta_c = \frac{W}{EI}\left(\frac{ba^3}{6l} - \frac{ab(l + b)a}{l}\right)$$

$$\underline{\delta_c = -\frac{Wa^2b^2}{3EIl}} \qquad (5.27)$$

When $a = b = l/2$ in equation (5.27),

$$\underline{\delta_c = -Wl^3/(48EI)}, \text{ as required.}$$

5.9.1   EXAMPLE 5.6   SIMPLY-SUPPORTED BEAM WITH COMPLEX LOAD

Determine an expression for the deflection distribution for the simply-supported beam of Fig. 5.6. Hence, of otherwise, obtain the position and value of the maximum deflection, given the following:

$$E = 2 \times 10^{11} \text{ N/m}^2 \qquad I = 2 \times 10^{-8} \text{ m}^4$$

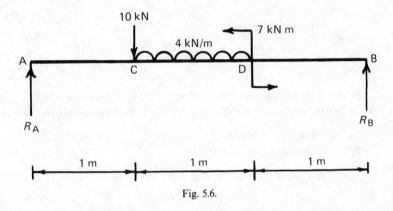

Fig. 5.6.

5.9.2

First, it is necessary to determine R_A, which can be obtained by taking moments about the point "B".

$$R_A \times 3 = 10 \times 2 + 4 \times 1 \times 1.5 + 7$$

$$\underline{R_A = 11 \text{ kN}}$$

In applying Macaulay's method to this beam, and remembering that the negative terms inside the Macaulay brackets must be ignored, it is necessary to make the distributed load of Fig. 5.6 equivalent to that of Fig. 5.7, which is essentially the same as that of Fig. 5.6.

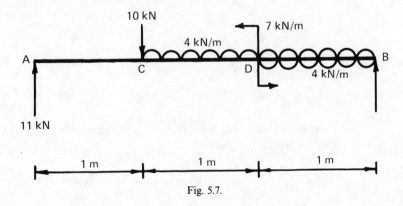

Fig. 5.7.

As the bending moment expression is different for sections AC, CD and DB, it will be necessary to apply Macaulay's method to each of these sections, as follows:

$$EI\frac{d^2y}{dx^2} = 11x \quad \bigg| \quad -10[x-1] \quad \bigg| \quad +\tfrac{4}{2}[x-2]^2$$
$$\bigg| \quad -\tfrac{4}{2}[x-1]^2 \quad \bigg| \quad -7[x-2]^0 \quad (5.28)$$

$$EI\frac{dy}{dx} = \frac{11x^2}{2} + A \quad \bigg| \quad -\tfrac{10}{2}[x-1]^2 \quad \bigg| \quad +\tfrac{2}{3}[x-2]^3$$
$$\bigg| \quad -\tfrac{2}{3}[x-1]^3 \quad \bigg| \quad -7[x-2] \quad (5.29)$$

$$EIy = \frac{11x^3}{6} + Ax + B \quad \bigg| \quad -\tfrac{5}{3}[x-1]^3 \quad \bigg| \quad +\frac{[x-2]^4}{6}$$
$$\bigg| \quad -\frac{[x-1]^4}{6} \quad \bigg| \quad -\tfrac{7}{2}[x-2]^2 \quad (5.30)$$

N.B. The Macaulay bracket for the couple must be written as in equation (5.28), so that integration can be carried out as in equations (5.29) and (5.30).

5.9.3 Boundary Conditions

A suitable boundary condition is as follows:

At $x = 0, y = 0$

therefore

$$\underline{B = 0}$$

N.B. As the terms in the Macaulay brackets in the second and third columns are negative, they must be ignored when applying the above boundary condition to equation (5.30).

Another suitable boundary condition is as follows:

At $x = 3, y = 0$

therefore

$$\underline{A = -10.056}$$

Substituting the above boundary conditions into equation (5.30), the following is obtained for the deflection y at any point x along the length of the beam

$$EIy = \frac{11x^3}{6}$$

$$-10.056x \quad \bigg| \quad -\tfrac{5}{3}[x-1]^3 - \frac{[x-1]^4}{6} \quad \bigg| \quad +\frac{[x-2]^4}{6}$$
$$-\tfrac{7}{2}[x-2]^2$$

The maximum deflection may occur in the span CD, where the condition $dy/dx = 0$ must be satisfied, i.e.

$$EI\frac{dy}{dx} = \frac{11x^2}{2} - 10.056 - 5[x-1]^2 - \tfrac{2}{3}[x-1]^3$$

or,

$$0 = 5.5x^2 - 10.056 - 5(x^2 - 2x + 1) - \tfrac{2}{3}(x^3 - 3x^2 + 3x - 1)$$

or,

$$- 0.667x^3 + 2.5x^2 + 8x - 14.389 = 0$$

which has three real roots, as follows:

$$x_1 = -2.913 \text{ m}$$

$$x_2 = 5.252 \text{ m}$$

$$x_3 = 1.411 \text{ m}$$

It is evident that the root of interest is $x_3 = 1.411$ m, as this is the only one that applies within the span CD, i.e.

$$\delta = \text{maximum deflection}$$

$$= \frac{1}{2 \times 10^{11} \times 2 \times 10^{-8}}$$

$$\times \left(\frac{11 \times 1.411^3}{6} - 10.05 \times 1.411 - \tfrac{5}{3}(0.411)^3 - \frac{0.411^4}{6} \right)$$

$$\delta = -2.288 \text{ mm}$$

5.10.1   STATICALLY INDETERMINATE BEAMS

So far, the beams that have been analysed were statically determinate, that is, their reactions and bending moments were determined solely from statical considerations.

For statically indeterminate beams, their analysis is more difficult, as their reactions and bending moments cannot be obtained from statical considerations alone. To demonstrate the method of analysing statically indeterminate beams, the following two simple cases will be considered.

5.11.1   EXAMPLE 5.7   ENCASTRÉ BEAM WITH UDL

Determine the end fixing moments M_F, and the maximum deflection for the encastré beam of Fig. 5.8.

$$EI \frac{d^2y}{dx^2} = -M_F + \frac{wl}{2}x - \frac{wx^2}{2}$$

$$EI \frac{dy}{dx} = -M_Fx + \frac{wlx^2}{4} - \frac{wx^3}{6} + A \tag{5.31}$$

At $x = 0, \dfrac{dy}{dx} = 0;$ therefore $\underline{A = 0}$

At $x = l, \dfrac{dy}{dx} = 0;$ therefore $0 = -M_Fl + \dfrac{wl^3}{4} - \dfrac{wl^3}{6}$

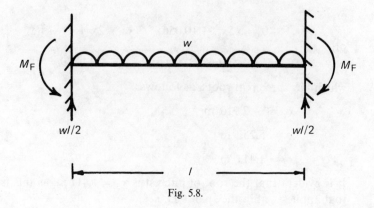

Fig. 5.8.

or,

$$M_F = \frac{wl^2}{12} \qquad\qquad (5.32)$$

i.e. the end fixing moment $M_F = wl^2/12$.

On integrating equation (5.31),

$$EIy = -\frac{M_F x^2}{2} + \frac{wlx^3}{12} - \frac{wx^4}{24} + B$$

At $\quad x = 0, y = 0;\quad$ therefore $\quad \underline{B = 0}$

i.e.

$$y = \frac{w}{EI}\left(-\frac{l^2 x^2}{24} + \frac{lx^3}{12} - \frac{x^4}{24}\right)$$

By inspection, the maximum deflection δ occurs at $x = l/2$, where,

$$\delta = -\frac{wl^4}{384EI} \qquad\qquad (5.33)$$

Equation (5.33) shows that the central deflection of an encastré beam is only one-fifth of that of the simply-supported case.

5.12.1   EXAMPLE 5.8   ENCASTRÉ BEAM WITH CONCENTRATED LOAD

Determine the end fixing moments and reactions and the deflection under the load for the encastré beam of Fig. 5.9.

$$EI\,\frac{d^2 y}{dx^2} = -M_A + R_A x \qquad\qquad\quad\bigg|\quad -W[x-a] \qquad (5.34)$$

$$EI\,\frac{dy}{dx} = -M_A x + \frac{R_A x^2}{2} + A \qquad\bigg|\quad -\frac{W}{2}[x-a]^2 \qquad (5.35)$$

$$EIy = -\frac{M_A x^2}{2} + \frac{R_A x^3}{6} + Ax + B \quad\bigg|\quad -\frac{W}{6}[x-a]^3 \qquad (5.36)$$

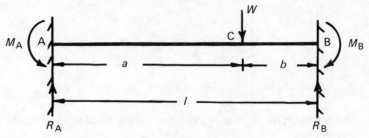

Fig. 5.9.

To determine A, B, R_A and M_A, it will be necessary to apply four boundary conditions to equations (5.35) and (5.36), as follows:

At $x = 0, y = 0$;   therefore $\underline{B = 0}$

At $x = 0, \dfrac{dy}{dx} = 0$;   therefore $\underline{A = 0}$

At $x = l, \dfrac{dy}{dx} = 0$ and $y = 0$

therefore

$$0 = -M_A l + \frac{R_A l^2}{2} - \frac{Wb^2}{2} \tag{5.37}$$

and,

$$0 = -\frac{M_A l^2}{2} + \frac{R_A l^3}{6} - \frac{Wb^3}{6} \tag{5.38}$$

From equation (5.37),

$$-M_A = -\frac{R_A l}{2} + \frac{Wb^2}{2l} \tag{5.39}$$

and from equation (5.38),

$$-M_A = -\frac{R_A l}{3} + \frac{Wb^3}{3l^2} \tag{5.40}$$

Equating (5.39) and (5.40),

$$-\frac{R_A l}{2} + \frac{Wb^2}{2l} = -\frac{R_A l}{3} + \frac{Wb^3}{3l^2}$$

or,

$$\underline{R_A = \frac{Wb^2}{l^3}\,(l + 2a)} \tag{5.41}$$

and from (5.39),

$$M_A = \frac{Wb^2}{2l^2}(l + 2a) - \frac{Wb^2}{2l}$$

$$M_A = \frac{Wab^2}{l^2} \tag{5.42}$$

Expressions for R_B and M_B can now be obtained from statical considerations, as follows.

Resolving vertically

$$R_A + R_B = W$$

therefore

$$R_B = \frac{Wa^2(l + 2b)}{l^3} \tag{5.43}$$

and,

$$M_B = \frac{Wa^2b}{l^2} \tag{5.44}$$

Substituting equations (5.41) and (5.42) into equation (5.36), with a value of $x = a$,

$$\delta_c = -\frac{Wa^3b^3}{3EIl^3} \tag{5.45}$$

If the beam of Fig. 5.9 were loaded symmetrically, so that $a = b = l/2$, then,

$$M_A = M_B = Wl/8 \tag{5.46}$$

and,

$$\delta_c = -Wl^3/(192EI) \tag{5.47}$$

From equation (5.47), it can be seen that the central deflection for a centrally loaded beam, with encastré ends, is one-quarter of the value for a similar beam with simply-supported ends.

5.13.1 MOMENT-AREA METHOD

This method is particularly useful if there is step variation with the sectional properties of the beam.

5.13.2

Now,

$$EI\frac{d^2y}{dx^2} = M$$

or,

$$\frac{\mathrm{d}^2 y}{\mathrm{d}x^2} = \frac{M}{EI} \tag{5.48}$$

Consider the deformed beam of Fig. 5.10, and apply equation (5.48) between the points A and B.

$$\int_{x_A}^{x_B} \frac{\mathrm{d}^2 y}{\mathrm{d}x^2}\,\mathrm{d}x = \int_{x_A}^{x_B} \left(\frac{M}{EI}\right)\mathrm{d}x$$

$$\left\{\frac{\mathrm{d}y}{\mathrm{d}x}\right\}_{x_A}^{x_B} = \int_{x_A}^{x_B} \left(\frac{M}{EI}\right)\mathrm{d}x$$

i.e.

$$\theta_B - \theta_A = \int_{x_A}^{x_B} \left(\frac{M}{EI}\right)\mathrm{d}x \tag{5.49}$$

$$= \text{area of } (M/EI) \text{ between } x_A \text{ and } x_B,$$

and if,

$$EI = \text{constant}$$

$$\theta_B - \theta_A = \frac{1}{EI} * \text{area of bending moment diagram between}$$

$$x_A \text{ and } x_B.$$

Furthermore, it can be seen that if both sides of equation (5.48) were multiplied by x, then

$$x\,\frac{\mathrm{d}^2 y}{\mathrm{d}x^2} = \frac{M}{EI}\,x$$

or,

$$\int_{x_A}^{x_B} x\,\frac{\mathrm{d}^2 y}{\mathrm{d}x^2}\,\mathrm{d}x = \int_{x_A}^{x_B} \frac{M}{EI}\,x\,\mathrm{d}x$$

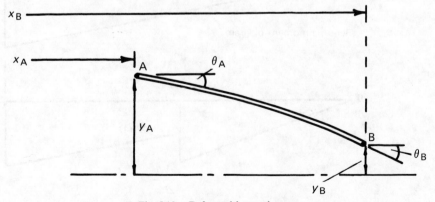

Fig. 5.10. Deformed beam element.

or,

$$\left\{ x \frac{dy}{dx} - y \right\}_{x_A}^{x_B} = \int_{x_A}^{x_B} \left(\frac{M}{EI} \right) x \, dx \qquad (5.50)$$

Suitable use of equations (5.49) and (5.50) can be used to solve many problems, and if EI is constant, then,

$$\left\{ x \frac{dy}{dx} - y \right\}_{x_A}^{x_B} = \frac{1}{EI} * \text{moment of area of the bending moment}$$
$$\text{diagram about the point "A".}$$

5.14.1   EXAMPLE 5.9   CANTILEVER WITH VARYING SECTION

Determine the end of deflection for the cantilever loaded with a point load at its free end, as shown in Fig. 5.11.

Take the point "B" to be at the built-in end and the point "A" to be at the free end.

From equation (5.50):

$$\left\{ x \frac{dy}{dx} - y \right\}_{0}^{l} = \frac{1}{E} * \text{moment of area of the } M/I \text{ diagram about "A".}$$

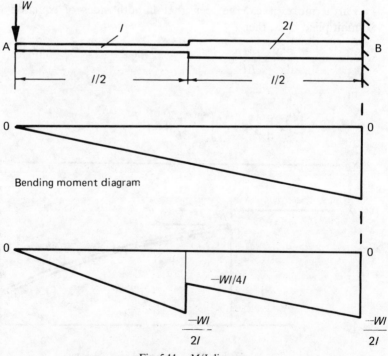

Fig. 5.11.   *M/I* diagram.

Therefore

$$\{(l\theta_B - y_B) - (0 * \theta_A - y_A)\}$$

$$= -\frac{1}{E}\left\{\frac{Wl}{2I} * \frac{3}{3} * \frac{l}{2} * \frac{l}{4} + \frac{Wl}{4I} * \frac{3l}{4} * \frac{l}{2}\right.$$

$$\left. + \frac{Wl}{4I} * \frac{l}{4} * \left(\frac{l}{4} + \tfrac{2}{3} * \frac{l}{2}\right)\right\}$$

but,

$$\theta_B = y_B = 0$$

therefore

$$y_A = -\frac{Wl^3}{EI}\left(\tfrac{1}{24} + \tfrac{3}{32} + \tfrac{5}{96}\right)$$

i.e.

$$\delta = y_A = -\frac{9Wl^3}{48EI} = -\frac{3Wl^3}{16EI}$$

If I were constant throughout the length of the cantilever,

$$\delta = -\frac{Wl^3}{96EI}\{4 + 18 + 10\}$$

$$\delta = -\frac{Wl^3}{3EI}$$

5.15.1

To solve beam problems, other than statically determinate ones or the simpler cases of statically indeterminate ones, is difficult and cumbersome, and for such cases, it is better to use computer methods [1, 5, 6]. Some of these computer methods are based on the *slope–deflection equations*, which are now derived.

5.16.1 SLOPE–DEFLECTION EQUATIONS

This is a boundary value problem and is dependent on the displacement boundary conditions of Fig. 5.12, where

$Y_1 = $ vertical force at node 1
$Y_2 = $ vertical force at node 2
$M_1 = $ clockwise moment at node 1
$M_2 = $ clockwise moment at node 2
$y_1 = $ vertical deflection at node 1
$y_2 = $ vertical deflection at node 2
$\theta_1 = $ rotation at node 1 (clockwise)
$\theta_2 = $ rotation at node 2 (clockwise)

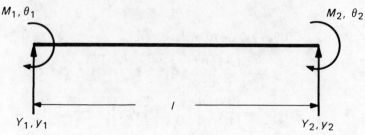

Fig. 5.12.

N.B. A node is defined as a point.
Resolving vertically

$$Y_1 = -Y_2 \tag{5.51}$$

Taking moments about the right end,

$$Y_1 = -(M_1 + M_2)/l \tag{5.52}$$

and,

$$Y_2 = (M_1 + M_2)/l$$

From equation (5.3),

$$EI\frac{\mathrm{d}^2 y}{\mathrm{d}x_2} = Y_1 x + M_1$$

$$= -(M_1 + M_2)x/l + M_1$$

$$EI\frac{\mathrm{d}y}{\mathrm{d}x} = -\frac{(M_1 + M_2)x^2}{2l} + M_1 x + A$$

At $\quad x = 0, \dfrac{\mathrm{d}y}{\mathrm{d}x} = -\theta_1$

therefore

$$A = -EI\theta_1 \tag{5.53}$$

therefore

$$EI\frac{\mathrm{d}y}{\mathrm{d}x} = -\frac{(M_1 + M_2)x^2}{2l} + M_1 x - EI\theta_1$$

and,

$$EIy = -\frac{(M_1 + M_2)x^3}{6l} + \frac{M_1 x^2}{2} - EI\theta_1 x + B$$

<u>At $\quad x = 0, y = y_1$;</u>   therefore   <u>$B = EIy_1$</u> $\tag{5.54}$

<u>At $\quad x = l, \dfrac{\mathrm{d}y}{\mathrm{d}x} = -\theta_2$</u>

therefore

$$- EI\theta_2 = - \frac{(M_1 + M_2)l}{2} + M_1 l - EI\theta_1 \qquad (5.55)$$

<u>At $x = l, y = y_2$</u>

therefore

$$EIy_2 = - \frac{(M_1 + M_2)l^2}{6} + \frac{M_1 l}{2} - EI\theta_1 l + EIy_1 \qquad (5.56)$$

From equations (5.51) to (5.56), the *slope-deflection equations* are obtained, as follows:

$$M_1 = \frac{4EI\theta_1}{l} + \frac{2EI\theta_2}{l} - \frac{6EI}{l^2} (y_1 - y_2)$$

$$Y_1 = - \frac{6EI\theta_1}{l^2} - \frac{6EI\theta_2}{l^2} + \frac{12EI}{l^3} (y_1 - y_2) \qquad (5.57)$$

$$M_2 = \frac{2EI\theta_1}{l} + \frac{4EI\theta_2}{l} - \frac{6EI}{l^2} (y_1 - y_2)$$

$$Y_2 = \frac{6EI\theta_1}{l^2} + \frac{6EI\theta_2}{l^2} - \frac{12EI}{l^3} (y_1 - y_2)$$

Equations (5.57) lend themselves to satisfactory computer analysis and form the basis of the finite element method, which is beyond the scope of the present book, but is described in detail in a number of other texts [5–7].

EXAMPLES FOR PRACTICE 5

1. Obtain an expression for the deflection y at any distance x from the left end of the uniform section beam of Fig. Q.5.1.

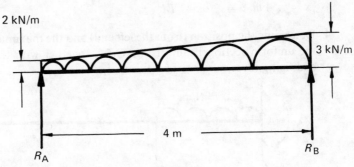

Fig. Q.5.1.

$$\left\{ y = \frac{1}{EI} (0.778x^3 - 0.0833x^4 - 2.083 * 10^{-3}x^5 - 6.579x) \right\}$$

2. Determine the value of the reactions and end fixing moments for the uniform section beam of Fig. Q.5.2. Hence, or otherwise, obtain an expression for the deflection y at any distance x from the left end of the beam.

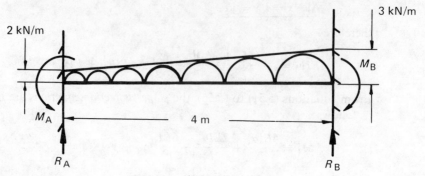

Fig. Q.5.2.

$$\left\{ R_A = 4.6 \text{ kN}, \ M_A = 3.2 \text{ kN m}, \ R_B = 5.4 \text{ kN}, \ M_B = 3.467 \text{ kN m}; \right.$$

$$\left. y = \frac{1}{EI} (0.777x^3 - 0.0833x^4 - 2.083 * 10^{-3}x^5 - 1.6x^2) \right\}$$

3. Determine the position (from the left end) and the value of the maximum deflection for the uniform section beam of Fig. Q.5.3.

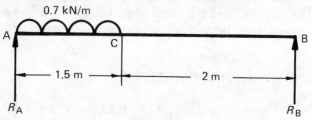

Fig. Q.5.3.

$$\{x = 1.54 \text{ m}, \ \delta = -0.543/EI\}$$

4. Determine the position (from the left end) and the maximum deflection for the uniform section beam of Fig. Q.5.4, together with the reactions and end fixing forces.

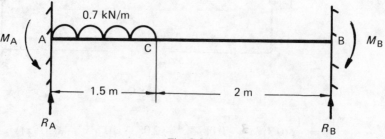

Fig. Q.5.4.

$\{R_A = 0.897$ kN, $R_B = 0.153$ kN, $M_A = 0.407$ kN m, $M_B = 0.155$ kN m; $x = 1.47$ m, $\delta = -0.103/EI\}$

5. Determine the deflections at the points C and D for the uniform section beam of Fig. Q.5.5, given that $EI = 4300$ kN m^2.
   {Portsmouth Polytechnic, June 1982}

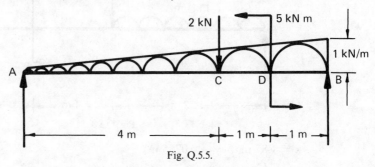

Fig. Q.5.5.

$\{\delta_C = -5.58 \times 10^{-3}$ m, $\delta_D = -3.32 \times 10^{-3}$ m$\}$

6. Determine the end fixing moments and reactions for the encastré beam shown in Fig. Q.5.6. The beam may be assumed to be of uniform section.

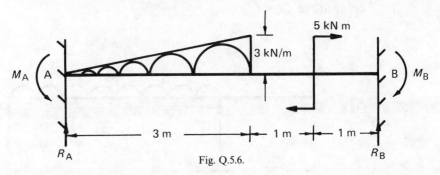

Fig. Q.5.6.

$\{R_A = 1.888$ kN, $R_B = 2.612$ kN, $M_A = 1.444$ kN m, $M_B = 0.506$ kN m$\}$

7. Determine the end fixing moments and reactions for the uniform section encastré beam of Fig. Q.5.7.

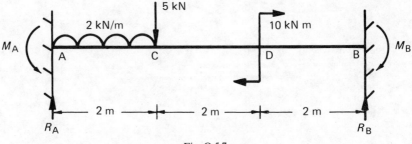

Fig. Q.5.7.

$\{R_A = 5.11$ kN, $R_B = 3.89$ kN, $M_A = 3.556$ kN m, $M_B = 2.889$ kN m$\}$

8. Determine the position and value of the maximum deflection of the simply-supported beam shown in Fig. Q.5.8, given that $EI = 100$ kN m^2. {Portsmouth Polytechnic, March 1982}.

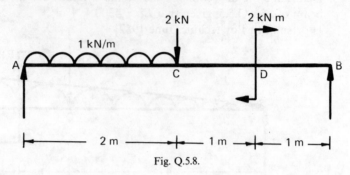

Fig. Q.5.8.

$\{x = 1.87$ m, $\delta = -0.0285$ m$\}$

9. The beam CAB is simply-supported at the points A and B and is subjected to a concentrated load W at the point C, together with a uniformly distributed load w, between the points A and B, as shown in Fig. Q.5.9.

   Determine the relationship between W and wl, so that no deflection will occur at the point C.

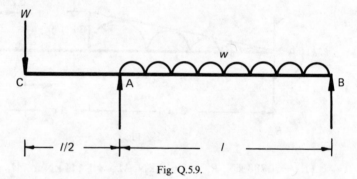

Fig. Q.5.9.

$\{w = 6W/l\}$

6

Torsion of Circular Sections

6.1.1

In engineering, it is often required to transmit power via a circular section shaft, and some typical examples of this are given below:

(a) Propulsion of a ship or a boat by a screw propeller, via a shaft.
(b) Transmission of power to the rear wheels of an automobile, via a shaft.
(c) Transmission of power from an electric motor to various types of machinery, via a shaft.

6.2.1 TORQUE (T)

A torque is defined as a twisting moment that acts on the shaft in an axial direction, as shown in Fig. 6.1, where T is according to the right-hand screw rule. This torque causes the end "B" to rotate by an angle θ, relative to "A", where,

$$\theta = \text{angle of twist}$$

6.2.2

Assuming that no other forces act on the shaft, the effect of T will be to cause the shaft to be subjected to a system of shearing stresses, as shown in Fig. 6.2. The system of shearing stresses acting on an element of the shaft of Fig. 6.2 is known as *pure shear*, as these shearing stresses are unaccompanied by direct stresses.

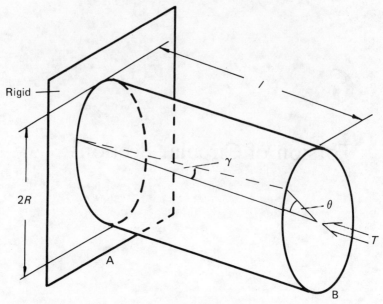

Fig. 6.1. Shaft under torque.

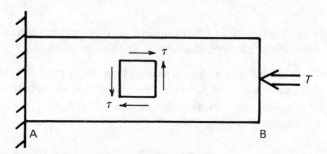

Fig. 6.2. Shaft in pure shear.

Later, it will be proven that, providing the angle of twist is small and the limit of proportionality is not exceeded, the shearing stresses τ will have a maximum value on the external surface of the shaft and their magnitude will decrease linearly to zero at the centre of the shaft.

6.3.1 ASSUMPTIONS MADE

The following assumptions are made in the theory used in this chapter:

(a) The shaft is of circular cross-section.
(b) The cross-section of the shaft is uniform throughout its length.
(c) The shaft is straight.
(d) The material is homogeneous, isotropic and obeys Hooke's law.
(e) Rotations are small and the limit of proportionality is not exceeded.
(f) Plane cross-sections remain plane during twisting.
(g) Radial lines across the shaft's cross-section remain radial during twisting.

6.4.1   Proof of $\dfrac{\tau}{r} = \dfrac{T}{J} = \dfrac{G\theta}{l}$

The following relationships, which are used in the torsional theory of *circular section shafts*, will now be proven.

$$\frac{\tau}{r} = \frac{T}{J} = \frac{G\theta}{l}$$

where,

> τ = shearing stress at any radius r
> T = applied torque
> J = polar second moment of area
> G = rigidity or shear modulus
> θ = angle of twist over a length l

6.4.2

From Fig. 6.1, it can be seen that

> γ = shear strain

and that,

> $\gamma l = R\theta$, providing θ is small

or,

$$\left(\frac{\tau}{G}\right)l = R\theta$$

therefore

$$\frac{\tau}{R} = \frac{G\theta}{l}$$

If the radial lines across the section remain radial on twisting, then it follows that the shearing stress is proportional to any radius r, so that,

$$\frac{\tau}{r} = \frac{G\theta}{l} \tag{6.1}$$

Similarly,

$$\frac{\tau}{r} = G\,\frac{d\theta}{dx} \tag{6.2}$$

where,

> $\dfrac{d\theta}{dx}$ = the change of the angle of twist over a length "dx".

[handwritten annotations in right margin:]

where $d = 2r$

or $\dfrac{\pi D^4}{2}$

$J = \dfrac{\pi d^4}{32}$ — solid Section

$J = \dfrac{\pi (D_2^4 - D_1^4)}{32}$ hollow Section

6.4.3

Consider a cylindrical shell element of radius r and thickness "dr", as shown by the shaded area of Fig. 6.3. The shearing stresses acting on the cross-section of this cylindrical shell are shown in Fig. 6.4, where they can be seen to act tangentially to the cross-section of the cylindrical shell.

From Fig. 6.4, it can be seen that these shearing stresses cause an elemental torque, δT, where,

$$\delta T = \tau * 2\pi r \, dr * r$$

but the total torque T is the sum of all the elemental torques acting on the section, i.e.

$$T = \sum \delta T$$

$$= 2\pi \int r^2 \tau \, dr \tag{6.3}$$

Substituting τ from equation (6.1) into equation (6.3):

$$T = 2\pi \int_0^R \frac{G\theta}{l} r^3 \, dr$$

$$= \frac{G\theta}{l} \left(\frac{\pi R^4}{2} \right)$$

but,

$$\frac{\pi R^4}{2} = J = \text{polar second moment of area of a circular section.}$$

Therefore

$$\frac{T}{J} = \frac{G\theta}{l} \tag{6.4}$$

From equations (6.1) and (6.4):

$$\frac{\tau}{r} = \frac{T}{J} = \frac{G\theta}{l} \tag{6.5}$$

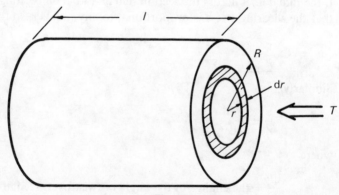

Fig. 6.3.

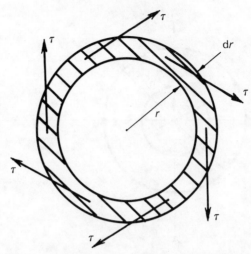

Fig. 6.4.   Shearing stresses acting on annular element.

From equation (6.5), it can be seen that τ is linearly proportional to the radius r, so that the shear stress at the centre is zero and it reaches a maximum value on the outermost surface of the shaft, as shown in Fig. 6.5.

6.4.4   Hollow Circular Sections

Equation (6.5) is also applicable to circular section tubes of uniform thickness, except that

$$J = \pi(R_2^4 - R_1^4)/2 = \text{polar second moment of area of an annulus}$$

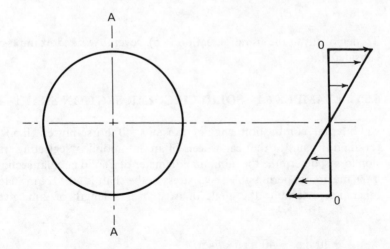

(a) Section   (b) Shear stress distribution at A–A

Fig. 6.5.   Distribution of τ.

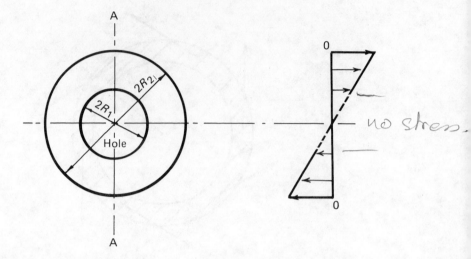

(a) Section   (b) Distribution of τ across A-A

Fig. 6.6.   Distribution of τ across a hollow circular section.

where,

R_2 = external radius of tube
R_1 = internal radius of tube

Using the above value of J, equation (6.5) can be applied to a circular section tube, between the radii R_1 and R_2, and the shear stress distribution will have the form shown of Fig. 6.6.

6.4.5

The *torsional stiffness of a shaft* =

$$k = GJ/l \tag{6.6}$$

To demonstrate the use of equation (6.5), several worked examples will be considered.

6.5.1   EXAMPLE 6.1   SOLID CIRCULAR SECTION SHAFT

An internal combustion engine transmits 40 horse-power (h.p.) at 200 revs/min (r.p.m.) to the rear wheels of an automobile. Neglecting transmission losses, determine the minimum diameter of a solid circular section shaft, if the maximum permissible shear stress in the shaft is 60 MN/m². Hence, or otherwise, determine the angle of twist over a length of 2 m, given that $G = 7.7 \times 10^{10}$ N/m².

$$40 \text{ h.p.} = 40 \text{ h.p.} \times 745.7 \frac{W}{\text{h.p.}}$$

$$= 29.83 \text{ kW}$$

Now,

$$\text{Power} = T\omega \tag{6.7}$$

where, Power is in units of watts:

$T = \text{torque (N m)}$

$\omega = \text{speed of rotation (rads/s)}$

In this case,

$$\omega = 2\pi \ \frac{\text{rads}}{\text{rev}} \times 200 \ \frac{\text{revs}}{\text{min}} \times \frac{1 \ \text{min}}{60 \ \text{s}}$$

$$\underline{\omega = 20.94 \ \text{rad/s}}$$

i.e.

$$29.83 \times 10^3 = T \times 20.94$$

therefore

$$\underline{T = 1.424 \ \text{kN m}} \tag{6.8}$$

Now the maximum permissible shearing stress $= 60 \ \text{MN/m}^2$, which occurs at the external radius R. Hence, from $\tau/r = T/J$

$$\frac{60 \times 10^6}{R} = \frac{1.424 \times 10^3}{(\pi R^4/2)}$$

$$R^3 = \frac{1.424 \times 10^3 \times 2}{\pi \times 60 \times 10^6}$$

$$\underline{R = 0.0247 \ \text{m}}$$

i.e.

$$\underline{\text{Shaft diameter} = 4.94 \ \text{cm}}$$

From,

$$\frac{\tau}{r} = \frac{G\theta}{l}$$

$$\theta = \frac{\tau l}{GR} = \frac{60 \times 10^6 \times 2}{7.7 \times 10^{10} \times 0.0247}$$

$$\underline{\theta = 0.0631 \ \text{rads} = 3.62°}$$

6.6.1   EXAMPLE 6.2   HOLLOW VERSUS SOLID SECTION SHAFT

If the shaft of Example 6.1 were in the form of a circular section tube, where the external diameter had twice the magnitude of the internal diameter, what would be the weight saving, if the hollow shaft were adopted in place of the solid one, assuming that both shafts had the same maximum shearing stress? What would be the resulting angle of twist in the hollow shaft?

Let

$$R = \text{external radius of hollow shaft}$$

therefore

$$J = \pi[R^4 - (R/2)^4]/2$$

$$\underline{J = 1.473\ R^4}$$

From,

$$\frac{\tau}{r} = \frac{T}{J}$$

$$\frac{60 \times 10^6}{R} = \frac{1.424 \times 10^3}{1.473\ R^4}$$

$$R^3 = \frac{1.424 \times 10^3}{1.473 \times 60 \times 10^6}$$

$$\underline{R = 0.0253\ \text{m}}$$

i.e.

External diameter of hollow shaft = 5.05 cm

Internal diameter of hollow shaft = 2.53 cm

Weight of solid shaft = $\rho g * \pi * 0.0247^2 * l$

Weight of hollow shaft = $\rho g * \pi[R^2 - (R/2)^2] * l$

therefore percentage weight saving = $\dfrac{0.0247^2 - (0.0253^2 - 0.01265^2)}{0.0247^2} \times 100$

$$= \underline{21.3}$$

From,

$$\frac{\tau}{r} = \frac{G\theta}{l}$$

$$\theta = \frac{60 \times 10^6 \times 2}{0.0253 \times 7.7 \times 10^{10}}$$

$$\underline{\theta = 0.0616\ \text{rads} = 3.53°}$$

6.7.1 FLANGED COUPLINGS

In some engineering situations, it is necessary to join together two shafts made from dissimilar materials. In such cases, it is usually undesirable to weld together the two shafts, and one method of overcoming this problem is by the

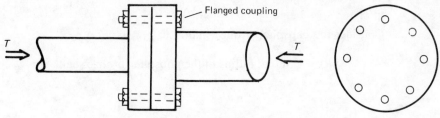

Fig. 6.7.   Flanged coupling.

use of a flanged coupling, as shown in Fig. 6.7. In this case the torque is transmitted from one shaft to the other by the action of shearing forces δF acting on the bolts, as shown in Fig. 6.8. Thus, if there are n bolts,

$$T = n * \delta F * R \tag{6.9}$$

where,

$$\delta F = \tau_b * \pi d^2/4 \tag{6.10}$$

R = pitch circle radius of bolt

τ_b = shearing stress in bolt

d = diameter of bolt

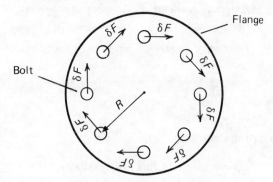

Fig. 6.8.   Shearing forces on bolts.

6.7.2   EXAMPLE 6.3   SHAFT WITH FLANGED COUPLING

A torque of 10 kN m is to be transmitted from a hollow phosphor bronze shaft, whose external diameter is twice its internal diameter, to a solid mild steel shaft, through a flanged coupling, with twelve bolts made from a high tensile steel.

Determine the dimensions of the shafts, the pitch circle diameter (PCD), and the diameters of the bolts, given that,

Maximum permissible shear stress in phosphor bronze = 20 MN m^2

Maximum permissible shear stress in mild steel = 30 MN/m^2

Maximum permissible stress in high tensile steel = 60 MN/m^2

6.7.3

Consider the phosphor bronze shaft. Let,

R_p = external radius of the phosphor bronze shaft

$$J = \frac{\pi[R_p^4 - (R_p/2)^4]}{2}$$

$$\underline{J = 1.473 \ R_p^4}$$

From,

$$\frac{\tau}{r} = \frac{T}{J}$$

$$R_p^3 = \frac{10 \times 10^3}{1.473 \times 20 \times 10^6}$$

$$\underline{R_p = 0.0698 \ m}$$

i.e.

External diameter of phosphor bronze shaft = 0.1396 m = 13.96 cm, say, 14 cm

Internal diameter of phosphor bronze shaft = 0.0698 m = 6.98 cm, say, 7 cm

6.7.4

Consider the steel shaft. Let,

R_s = external radius of steel shaft

$$J = \frac{\pi * R_s^4}{2} = 1.571 \ R_s^4$$

From,

$$\frac{\tau}{r} = \frac{T}{J}$$

$$R_s^3 = \frac{10 \times 10^3}{1.571 \times 30 \times 10^6}$$

$$\underline{R_s = 0.0596 \ m}$$

i.e.

Diameter of steel shaft = 0.1193 m = 11.9 cm, say, 12 cm

6.7.5 Bolts on Flanged Coupling

As the external diameter of the hollow shaft is 14 cm, it will be necessary to assume that the pitch circle diameter of the bolts on the flanged coupling is larger than this, so that the bolts can be accommodated.

Assumption

Let,

$$D = \text{pitch circle diameter}$$
$$= 20 \text{ cm (to allow for fitting)}$$
$$n = \text{number of bolts} = 12$$

From equation (6.9):

$$\delta F = \frac{T}{nR} = \frac{10 \text{ kN m}}{12 \times 0.1}$$

$$\underline{\delta F = 8.33 \text{ kN}}$$

From equation (6.10):

$$d^2 = \frac{4 * \delta F}{\pi * \tau_b} = \frac{4 \times 8.33 \times 10^3}{\pi \times 60 \times 10^6}$$

$$d = 0.0133 \text{ m}$$

$$\underline{d = 1.33 \text{ cm, say, 1.5 cm}}$$

6.8.1 KEYED COUPLINGS

Another method of transmitting power through shafts of dissimilar materials is through the use of keyed couplings, as shown in Fig. 6.9. In such cases, the key, which is the male portion of the coupling, is on the section of one shaft, and the keyway, which is the female portion of the coupling, is on the section of the other shaft. It is evident that the manufacture of the "key" and the "keyway" has to be precise, so that the former fits snugly into the latter. Precise analysis of keyed couplings is difficult and beyond the scope of this

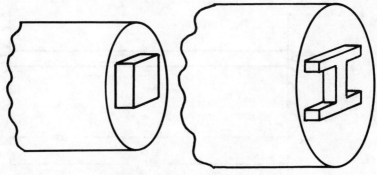

Fig. 6.9. Keys on shaft sections.

book, but an approximate design can be made by assuming that the shear stress distribution in a key varies linearly with the radial distance of a point on the key from the centre of the shaft. When such an approximate design is made, it should be ensured that the "factor of safety" is so large that the "factor of ignorance" is well contained.

6.9.1   EXAMPLE 6.4   SHAFT WITH INTERMEDIATE TORQUE

Determine the torque diagram for the shaft AB, which is subjected to an intermediate torque of magnitude T at the point C, as shown in Fig. 6.10. The shaft may be assumed to be firmly fixed at its ends. From Newton's third law of motion, it is evident that the intermediate torque T will be resisted by the end torques T_1 and T_2, acting in an opposite direction to T.

$$T_1 + T_2 = T \tag{6.11}$$

Let,

$$\theta_c = \text{the rotation at the point "C".}$$

Now, from

$$\frac{T}{J} = \frac{G\theta}{l}$$

$$T_1 = \frac{GJ\theta_c}{a} \tag{6.12}$$

and,

$$T_2 = \frac{GJ\theta_c}{b} \tag{6.13}$$

Dividing (6.12) by (6.13):

$$\frac{T_1}{T_2} = \frac{b}{a}$$

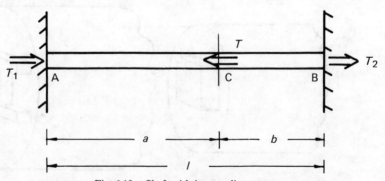

Fig. 6.10.   Shaft with intermediate torque.

or,

$$T_1 = bT_2/a \tag{6.14}$$

Substituting equation (6.14) into equation (6.11):

$$T_2 = \frac{T}{(1 + b/a)} = \frac{Ta}{l} \tag{6.15}$$

and,

$$T_1 = \frac{Tb}{l} \tag{6.16}$$

From equations (6.15) and (6.16), it can be seen that the torque diagram is as shown in Fig. 6.11.

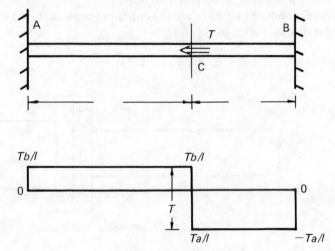

Fig. 6.11.   Torque diagram.

6.9.2

If the shaft of Fig. 6.10 were made from two different sections, joined together at "C", so that,

J_1 = polar 2nd moment of area of shaft AC
J_2 = polar 2nd moment CB of area
R_1 = external radius of shaft AC
R_2 = external radius CB of shaft
τ_1 = maximum shear stress in shaft AC
τ_2 = maximum shear stress in shaft CB

then,

$$\frac{\tau_1}{R_1} = \frac{T_1}{J_1} = \frac{G\theta_c}{a} \tag{6.17}$$

and,

$$\frac{\tau_2}{R_2} = \frac{T_2}{J_2} = \frac{G\theta_c}{b} \tag{6.18}$$

From equations (6.17) and (6.18), the two shafts can be designed.

6.10.1 COMPOUND SHAFTS

Composite shafts are made up from several smaller shafts with different material properties. The use of different materials for the manufacture of shafts is required when a shaft has to pass through different fluids, some of which are hostile to certain materials.

Composite shafts are usually either in series, as shown in Fig. 6.12, or in parallel, as shown in Fig. 6.13.

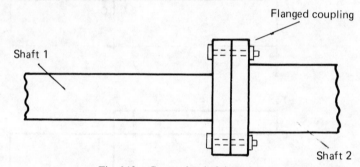

Fig. 6.12. Composite shaft in series.

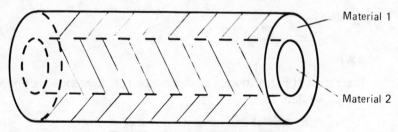

Fig. 6.13. Composite shafts in parallel.

6.11.1 EXAMPLE 6.5 COMPOUND SHAFT IN SERIES

The compound shaft ACB is fixed at the points A and B and subjected to a torque of 5 kN m at the point C, as shown in Fig. 6.14. If both parts of the shaft are of solid circular sections, determine the angle of twist at the point C

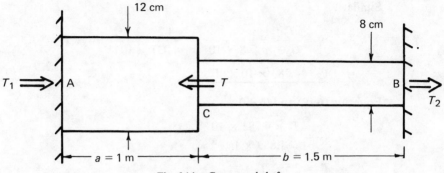

Fig. 6.14.   Compound shaft.

and the maximum shearing stresses in each section, assuming the following apply:

Shaft 1

$$G_1 = 2.6 \times 10^{10} \text{ N/m}^2$$
$$R_1 = 6 \text{ cm}$$
$$a = 1 \text{ m}$$

Shaft 2

$$G_2 = 7.8 \times 10^{10} \text{ N/m}^2$$
$$R_2 = 4 \text{ cm}$$
$$b = 2 \text{ m}$$

6.11.2

$$J_1 = \frac{\pi \times 0.06^4}{2} = 2.036 \times 10^{-5} \text{ m}^4$$

$$J_2 = \frac{\pi \times 0.04^4}{2} = 4.021 \times 10^{-6} \text{ m}^4$$

Let,

θ_c = the angle of twist at the point "C".

Now, from

$$\frac{T}{J} = \frac{G\theta}{l}$$

$$\theta_c = \frac{T_1 a}{G_1 J_1} = \frac{T_1 \times 1}{2.6 \times 10^{10} \times 2.036 \times 10^{-5}}$$

$$\underline{\theta_c = 1.889 \times 10^{-6} \, T_1} \tag{6.19}$$

Similarly,

$$\theta_c = \frac{T_2 b}{G_2 J_2} = \frac{T_2 \times 1.5}{7.8 \times 10^{10} \times 4.021 \times 10^{-6}}$$

$$\theta_c = 4.783 \times 10^{-6} \, T_2 \tag{6.20}$$

Equating (6.19) and (6.20):

$$T_1 = \frac{T_2 \times 4.783 \times 10^{-6}}{1.889 \times 10^{-6}}$$

$$\underline{T_1 = 2.532 \, T_2} \tag{6.21}$$

Now,

$$T = T_1 + T_2$$

$$5 \text{ kN m} = T_2(1 + 2.532)$$

$$\underline{T_2 = 1.416 \text{ kN m}}$$

$$\underline{T_1 = 3.584 \text{ kN m}}$$

From equation (6.20):

$$\theta_c = 4.783 \times 10^{-6} \times 1.416 \times 10^3$$

$$\underline{\underline{\theta_c = 6.773 \times 10^{-3} \text{ rads} = 0.388°}}$$

6.11.3

From,

$$\frac{\tau}{r} = \frac{T}{J}$$

$$\tau_1 = \frac{T_1 R_1}{J_1} = \frac{3.584 \times 10^3 \times 0.06}{2.036 \times 10^{-5}}$$

$$\underline{\tau_1 = 10.56 \text{ MN/m}^2}$$

$$\tau_2 = \frac{T_2 R_2}{J_2} = \frac{1.416 \times 10^3 \times 0.04}{4.021 \times 10^{-6}}$$

$$\underline{\tau_2 = 14.09 \text{ MN/m}^2}$$

6.12.1   EXAMPLE 6.6   COMPOUND SHAFT IN PARALLEL

The compound shaft of Fig. 6.15 is subjected to a torque of 5 kN m. Determine the angle of twist and the maximum shearing stresses in materials 1 and 2, given the following values for the moduli of rigidity:

$$G_1 = 2.5 \times 10^{10} \text{ N/m}^2 \text{—Material 1}$$

$$G_2 = 7.8 \times 10^{10} \text{ N/m}^2 \text{—Material 2}$$

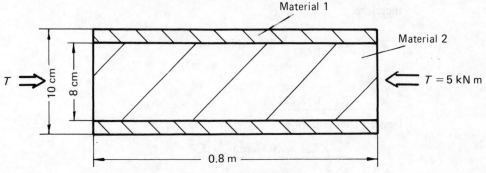

Fig. 6.15.   Compound shaft.

6.12.2   Assumptions Made

(a) No slipping takes place at the common interface.

(b) Radial lines remain straight on twisting.

From (b),

$$\underline{\theta_1 = \theta_2}$$

$$J_1 = \frac{\pi(0.05^4 - 0.04^4)}{2} = \underline{5.796 \times 10^{-6} \ \text{m}^4}$$

$$J_2 = \frac{\pi \times 0.04^4}{2} \qquad = \underline{4.021 \times 10^{-6} \ \text{m}^4}$$

N.B. Although the thickness of material 1 is only 1 cm, $J_1 > J_2$, which shows that a hollow shaft has a better strength: weight ratio than a solid one. From,

$$\frac{T}{J} = \frac{G\theta}{l}$$

$$\theta = \frac{T_1 l}{G_1 J_1} = \frac{T_1 \times 0.8}{2.5 \times 10^{10} \times 5.796 \times 10^{-6}}$$

$$\underline{\theta = 5.521 \times 10^{-6} \ T_1} \tag{6.22}$$

but,

$$\theta = \frac{T_2 l}{G_2 J_2} = \frac{T_2 \times 0.8}{7.8 \times 10^{10} \times 4.021 \times 10^{-6}}$$

$$\underline{\theta = 2.551 \times 10^{-6} \ T_2} \tag{6.23}$$

Equating (6.22) and (6.23):

$$\underline{T_1 = 0.462 \ T_2}$$

but,

$$T = T_1 + T_2$$

therefore

$$T_2 = \frac{5}{1.462} = 3.42 \text{ kN m}$$

and,

$$T_1 = 1.58 \text{ kN m}$$

Hence, from equation (6.23):

$$\theta = 8.724 \times 10^{-3} \text{ rads} = 0.5°$$

6.12.3

Let,

τ_1 = maximum shearing stress in material 1

$$= \frac{T_1 \times R_1}{J_1} = \frac{1.58 \times 10^3 \times 0.05}{5.796 \times 10^{-6}}$$

$$\tau_1 = 13.63 \text{ MN/m}^2$$

6.12.4

Let,

τ_2 = maximum shearing stress in material 2

$$= \frac{T_2 \times R_2}{J_2} = \frac{3.42 \times 10^3 \times 0.04}{4.021 \times 10^{-6}}$$

$$\tau = 34.02 \text{ MN/m}^2$$

N.B. From the practical point of view, if the shaft of Example 6.6 were used out of doors, in a *normal U.K. atmosphere*, convenient materials may have been aluminium alloy for material 1 and steel for material 2. The reasons for such a choice would be as follows:

(a) Both materials are relatively inexpensive.
(b) Mild steel is, in general, stronger and stiffer than aluminium alloy.
(c) Aluminium alloy does not rust; hence, its use as a sheath over a steel core.

To *manufacture* the compound shaft of Fig. 6.15, it is necessary to make the internal diameter of the steel shaft slightly larger than the external diameter of the aluminium alloy shaft and to "join" the two together by either heating the aluminium alloy shaft and/or cooling the steel shaft.

6.13.1 Tapered Shafts

There are a number of texts that treat the analysis of tapered shafts by using a simple extension of equation (6.5), but this method of analysis is incorrect, as the longitudinal shearing stresses produced on torsion are not parallel to the axis of the shaft.

For the analysis of tapered shafts, see reference [8].

6.14.1   CLOSE-COILED HELICAL SPRINGS

Equation (6.5) can be used for the stress analysis of close-coiled helical springs, providing the angle of helix and the deflections are small.

For the close-coiled helical spring of Fig. 6.16, most of the coils will be under torsion, where the torque T equals $WD/2$, so that the maximum deflection at B, caused by W, would be due to the combined effect of all the rotations of all the circular sections of the wire. Thus, the close-coiled helical spring can be assumed to be equivalent to a long shaft of length πDn, as shown in Fig. 6.17, where,

δ = deflection of spring at B due to W
D = mean coil diameter = $2R$
d = wire diameter

From Fig. 6.17,

$$\theta = \frac{T}{J} * \frac{\pi Dn}{G}$$

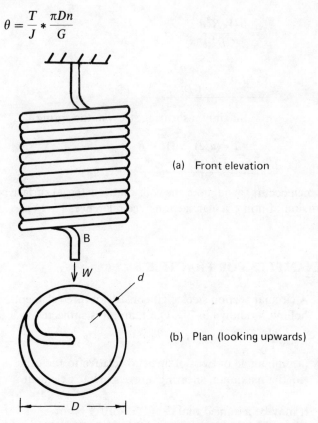

(a)   Front elevation

(b)   Plan (looking upwards)

Fig. 6.16.   Close-coiled helical spring.

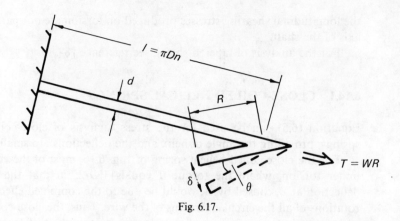

Fig. 6.17.

or,

$$\theta = \frac{W * R * \pi * 2Rn}{\pi * \dfrac{d^4}{32} * G}$$

$$\theta = \frac{64WR^2n}{GD^4}$$

$$\delta = R\theta = \frac{64WR^3n}{GD^4}$$

τ = maximum shearing stress in the spring

$$\tau = \frac{T * (d/2)}{\pi * (d^4/32)} = \frac{16WR}{\pi d^3}$$

which occurs throughout the coils on the external surface of the wire. For the torsion of non-circular sections, see references [8–11].

EXAMPLES FOR PRACTICE 6

1. A circular section steel shaft consists of three elements, two solid and one hollow, as shown in Fig. Q.6.1, and it is subjected to a torque of 3 kN m. Determine:

   (a) the angle of twist of one end relative to the other;
   (b) the maximum shearing stresses in each element.

   It may be assumed that $G = 7.7 \times 10^{10}$ N/m^2.
   {1.91°, 70.75 MN/m^2, 29.84 MN/m^2, 31.83 MN/m^2}

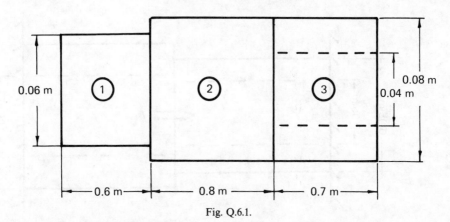

Fig. Q.6.1.

2. If the shaft of Example 1 were constructed from three separate materials, namely steel in element ①, aluminium alloy in element ② and manganese bronze in element ③, determine,

   (a) the total angle of twist;
   (b) the maximum shearing stresses in each element.

$$G_1 = 7.7 \times 10^{10} \text{ N/m}^2, \quad G_2 = 2.5 \times 10^{10} \text{ N/m}^2,$$
$$G_3 = 3.9 \times 10^{10} \text{ N/m}^2$$

{3.24°, 70.75 MN/m², 29.84 MN/m², 31.83 MN/m²}

3. Determine the output torque of an electric motor which outputs 5 kW at 25 revs/s.

   (a) If this torque is transmitted through a tube of external diameter 20 mm, determine the internal diameter, if the maximum permissible shearing stress in the shaft is 35 MN/m².
   (b) If this shaft is to be connected to another one, via a flanged coupling, determine a suitable bolt diameter, if four bolts are used on a pitch circle diameter of 30 mm, and the maximum permissible shearing stresses in the bolts equal 45 MN/m².

{31.83 N m; 16.11 mm, say, 16 mm; 3.88 mm, say, 4 mm}

4. A compound shaft consists of two equal length hollow shafts joined together in series and subjected to a torque *T*, as shown in Fig. Q.6.4.

   If the shaft on the left is made from steel, and the shaft on the right of the figure is made from aluminium alloy, determine the maximum permissible value of *T*, given the following:

For steel

$$G = 7.7 \times 10^{10} \text{ N/m}^2$$
Maximum permissible shear stress = 140 MN/m².

For aluminium alloy
$$G = 2.6 \times 10^{10} \text{ N/m}^2$$
Maximum permissible shear stress = 90 MN/m².

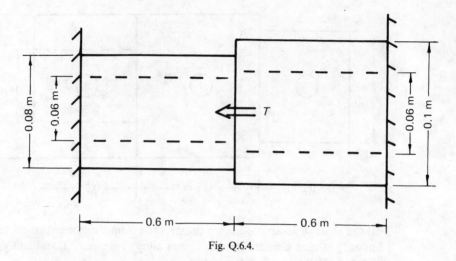

Fig. Q.6.4.

{Portsmouth Polytechnic, June 1980}

{19.72 kN m}

5. A compound shaft consists of a solid aluminium alloy cylinder of length 1 m and diameter 0.1 m, connected in series to a steel tube of the same length and external diameter, and of thickness 0.02 m.

    The shaft is fixed at its ends and it is subjected to an intermediate torque of 9 kN m at the joint.

    Determine the angle of twist and the maximum shear stress in the two halves.

$$G_{(steel)} = 7.7 \times 10^{10} \text{ N/m}^2$$
$$G_{(Al \text{ alloy})} = 2.6 \times 10^{10} \text{ N/m}^2$$

{Portsmouth Polytechnic, June 1982}

{$0.569°$, $\tau_{(al. \text{ alloy})} = 12.9 \text{ MN/m}^2$, $\tau_{(steel)} = 37.82 \text{ MN/m}^2$}

6. A compound shaft consists of two elements of equal length, joined together in series. If one element of the shaft is constructed from gunmetal tube and the other from solid steel, where the external diameter of the steel shaft and the internal diameter of the gunmetal shaft equal 50 mm, determine the external diameter of the gunmetal shaft, if the two shafts are to have the same torsional stiffness.

    Determine also, the maximum permissible torque that can be applied to the shaft, given the following:

For gunmetal

$$G = 3 \times 10^{10} \text{ N/m}^2$$
Maximum permissible shearing stress = 45 MN/m^2.

For steel

$$G = 7.5 \times 10^{10} \text{ N/m}^2$$
Maximum permissible shearing stress = 90 MN/m^2.

{Portsmouth Polytechnic, June 1977}

{68.4 mm, 2018 N m}

7. A compound shaft consists of a solid steel core, which is surrounded co-axially by an aluminium bronze sheath.

Determine suitable values for the diameters of the shafts if the steel core is to carry two-thirds of a total torque of 500 N m and that the limiting stresses are not exceeded.

For aluminium alloy
$$G = 3.8 \times 10^{10} \text{ N/m}^2$$
Maximum permissible shear stress = 18 MN/m^2.

For steel

$$G = 7.7 \times 10^{10} \text{ N/m}^2$$
Maximum permissible shear stress = 36 MN/m^2.

{45.4 mm, 38.1 mm, where the stress in the bronze shaft is the design criterion}

7

Complex Stress and Strain

7.1.1

The theory of Chapter 2 dealt with one-dimensional stress systems, and although this theory can be satisfactorily applied to a number of engineering problems, it breaks down when two- and three-dimensional stress systems are met.

In this chapter, we will concern ourselves mostly with two-dimensional stress systems, which usually occur with in-plane plate problems, such as met on the decks of a ship or the fuselage of an aircraft, and also in the torsion of circular section shafts.

The chapter, therefore, will commence with the mathematical theory of two-dimensional stress systems which, in general, is easier to understand than the mathematical theory of two-dimensional strain systems. After application of the mathematical theory for two-dimensional stress systems to a few practical examples, the mathematical theory of two-dimensional strain will be given, followed by engineering applications.

7.2.1 COMPLEX STRESS

A typical system of two-dimensional complex stresses, acting on an infinitesimally small rectangular lamina, is shown in Fig. 7.1, where the direct stresses σ_x and σ_y are accompanied by a set of shearing stresses τ_{xy}, acting in the x-y plane. These stresses are known as *co-ordinate stresses*, and Fig. 7.1 shows their positive signs, including those for the shear stress system.

The reason for choosing a lamina of rectangular shape is to simplify mathematical computation. That is, σ_x has no component in the y direction

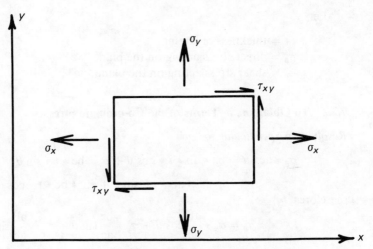

Fig. 7.1.   Complex stress system.

and σ_y has no component in the x direction, and also because, τ_{xy} is complementary and equal.

In Fig. 7.1, the existence of the direct stresses is self-evident, but the reader may not, as readily, accept the system of shearing stresses. These, however, can be explained with the aid of Section 2.3.2, where it can be seen that shearing stresses are complementary and equal; thus, for completeness, it is necessary to assume the shear stress system of Fig. 7.1.

In practice, however, it will be useful to have relationships of the direct and shearing stresses at any angle θ, in terms of the co-ordinate stresses.

Consider the stress system, acting on the sub-element "abc" of Fig. 7.2, where "ac" is at an angle θ to the y axis and σ_θ is at an angle θ (anti-clockwise) to the x axis.

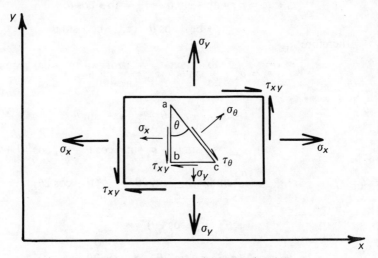

Fig. 7.2.   Stress system on sub-element "abc".

Let,

> t = thickness of lamina
> σ_θ = direct stress acting on the plane "ac"
> τ_θ = shear stress acting on the plane "ac"

7.2.2 To Obtain σ_θ in Terms of the Co-ordinate Stresses

Resolving perpendicular to "ac"

$$\sigma_\theta * ac * t = \sigma_x * ab * t * \cos\theta + \sigma_y * bc * t * \sin\theta$$
$$+ \tau_{xy} * ab * t * \sin\theta + \tau_{xy} * bc * t * \cos\theta$$

therefore

$$\sigma_\theta = \sigma_x * \frac{ab}{ac}\cos\theta + \sigma_y * \frac{bc}{ac}\sin\theta + \tau_{xy} * \frac{ab}{ac}\sin\theta$$

$$+ \tau_{xy} * \frac{bc}{ac}\cos\theta$$

$$= \sigma_x \cos^2\theta + \sigma_y \sin^2\theta + 2\tau_{xy}\sin\theta\cos\theta$$

$$= \frac{\sigma_x}{2}(1 + \cos 2\theta) + \frac{\sigma_y}{2}(1 - \cos 2\theta) + 2\tau_{xy}\sin\theta\cos\theta$$

therefore

$$\sigma_\theta = \tfrac{1}{2}(\sigma_x + \sigma_y) + \tfrac{1}{2}(\sigma_x - \sigma_y)\cos 2\theta + \tau_{xy}\sin 2\theta \tag{7.1}$$

7.2.3 To Determine τ_θ in Terms of the Co-ordinate Stresses

Resolving parallel to "ac"
N.B. The effects of t can be ignored, as t appears on both sides of the equation and therefore cancels out.

$$\tau_\theta * ac = \sigma_x * ab * \sin\theta - \tau_{xy} * ab * \cos\theta$$
$$- \sigma_y * bc * \cos\theta + \tau_{xy} * bc * \sin\theta$$

therefore

$$\tau_\theta = \sigma_x * \frac{ab}{ac}\sin\theta - \tau_{xy} * \frac{ab}{ac}\cos\theta$$

$$- \sigma_y * \frac{bc}{ac}\cos\theta + \tau_{xy} * \frac{bc}{ac}\sin\theta$$

$$= \sigma_x \cos\theta\sin\theta - \sigma_y \sin\theta\cos\theta - \tau_{xy}\cos^2\theta + \tau_{xy}\sin^2\theta$$

$$= \frac{\sigma_x}{2}\sin 2\theta - \frac{\sigma_y}{2}\sin 2\theta - \frac{\tau_{xy}}{2}(1 + \cos 2\theta)$$

$$+ \frac{\tau_{xy}}{2}(1 - \cos 2\theta)$$

$$\tau_\theta = \tfrac{1}{2}(\sigma_x - \sigma_y)\sin 2\theta - \tau_{xy}\cos 2\theta \tag{7.2}$$

One of the main reasons for obtaining equations (7.1) and (7.2) is to determine the magnitudes and directions of the maximum values of σ_θ and τ_θ.

7.3.1   PRINCIPAL STRESSES (σ_1 AND σ_2)

The maximum and minimum values of σ_θ occur when,

$$\frac{d\sigma_\theta}{d\theta} = 0$$

Hence, from equation (7.1),

$$-(\sigma_x - \sigma_y)\sin 2\theta + 2\tau_{xy}\cos 2\theta = 0$$

or,

$$\tan 2\theta = \frac{2\tau_{xy}}{(\sigma_x - \sigma_y)} \tag{7.3}$$

7.3.2   To Determine the Maximum and Minimum Values of σ_θ (i.e. $\hat{\sigma}_\theta$)

If equation (7.3) is represented by the mathematical triangle of Fig. 7.3, then,

$$\sin(2\theta) = \pm \frac{2\tau_{xy}}{\sqrt{[(\sigma_x - \sigma_y)^2 + 4\tau_{xy}^2]}} \tag{7.4}$$

and,

$$\cos(2\theta) = \pm \frac{(\sigma_x - \sigma_y)}{\sqrt{[(\sigma_x - \sigma_y)^2 + 4\tau_{xy}^2]}} \tag{7.5}$$

Substituting equations (7.4) and (7.5) into equation (7.1),

$$\hat{\sigma}_\theta = \tfrac{1}{2}(\sigma_x + \sigma_y) \pm \tfrac{1}{2}\frac{(\sigma_x - \sigma_y)(\sigma_x - \sigma_y)}{\sqrt{[(\sigma_x - \sigma_y)^2 + 4\tau_{xy}^2]}}$$

$$\pm \tau_{xy} \cdot \frac{2\tau_{xy}}{\sqrt{[(\sigma_x - \sigma_y)^2 + 4\tau_{xy}^2]}}$$

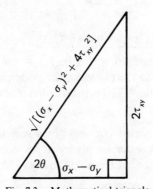

Fig. 7.3.   Mathematical triangle.

or,

$$\hat{\sigma}_\theta = \tfrac{1}{2}(\sigma_x + \sigma_y) \pm \tfrac{1}{2}\sqrt{[(\sigma_x - \sigma_y)^2 + 4\tau_{xy}^2]}$$

i.e. $\hat{\sigma}_\theta$ has two values: a maximum value σ_1 and a minimum value σ_2, where

$$\sigma_1 = \text{maximum principal stress}$$

$$= \tfrac{1}{2}(\sigma_x + \sigma_y) + \sqrt{\tfrac{1}{2}[(\sigma_x - \sigma_y)^2 + 4\tau_{xy}^2]} \qquad (7.6)$$

$$\sigma_2 = \text{maximum principal stress}$$

$$= \tfrac{1}{2}(\sigma_x + \sigma_y) - \tfrac{1}{2}\sqrt{[(\sigma_x - \sigma_y)^2 + 4\tau_{xy}^2]} \qquad (7.7)$$

N.B. Even though σ_2 is the minimum principal stress, if it is negative, it can be larger in magnitude than σ_1.

7.4.1 Maximum Shear Stress ($\hat{\tau}$)

$\hat{\tau}$ occurs when

$$\frac{d\tau_\theta}{d_\theta} = 0$$

i.e. from equation (7.2):

$$0 = \tfrac{1}{2}(\sigma_x - \sigma_y).\, 2\cos 2\theta + 2\tau_{xy}\sin 2\theta$$

$$\tan 2\theta = -\frac{(\sigma_x - \sigma_y)}{2\tau_{xy}} = \frac{(\sigma_y - \sigma_x)}{2\tau_{xy}} \qquad (7.8)$$

Equation (7.8) can be represented by the mathematical triangle of Fig. 7.4, where it can be seen that,

$$\cos 2\theta = \frac{2\tau_{xy}}{\sqrt{[(\sigma_x - \sigma_y)^2 + 4\tau_{xy}^2]}} \qquad (7.9)$$

$$\sin 2\theta = \frac{(\sigma_y - \sigma_x)}{\sqrt{[(\sigma_x - \sigma_y)^2 + 4\tau_{xy}^2]}} \qquad (7.10)$$

Substituting equations (7.9) and (7.10) into equation (7.2),

$$\hat{\tau} = \pm\tfrac{1}{2}\frac{(\sigma_x - \sigma_y)(\sigma_y - \sigma_x)}{\sqrt{[(\sigma_x - \sigma_y)^2 + 4\tau_{xy}^2]}} \pm \frac{\tau_{xy}\cdot 2\cdot\tau_{xy}}{\sqrt{[(\sigma_x - \sigma_y)^2 + 4\tau_{xy}^2]}}$$

$$\hat{\tau} = \pm[\tfrac{1}{4}(\sigma_x - \sigma_y)^2 + \tau_{xy}^2]^{1/2} \qquad (7.11)$$

If Fig. 7.4 is compared with Fig. 7.3, it can be seen that $\hat{\tau}$ occurs on planes which are at 45° to the planes of the principal stresses. Furthermore, it can be seen that if equation (7.3) is substituted into equation (7.2), then $\underline{\tau_\theta = 0}$ on the principal planes.

i.e. *There are no shearing stresses on a principal plane*, and that the maximum shearing stresses occur on planes at 45° to the principal planes.

It can also be seen from equations (7.6), (7.7) and (7.11), that,

$$\hat{\tau} = (\sigma_1 - \sigma_2)/2 \qquad (7.12)$$

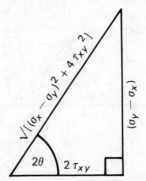

Fig. 7.4.   Mathematical triangle.

7.5.1 MOHR'S STRESS CIRCLE

Equations (7.1) and (7.2) can be represented in terms of principal stresses, by substituting

$$\sigma_x = \sigma_1$$

$$\sigma_y = \sigma_2 \quad \text{on a principal plane,}$$

and

$$\tau_{xy} = 0$$

so that,

$$\sigma_\theta = \tfrac{1}{2}(\sigma_1 + \sigma_2) + \tfrac{1}{2}(\sigma_1 - \sigma_2)\cos 2\theta \tag{7.13}$$

and

$$\tau_\theta = \tfrac{1}{2}(\sigma_1 - \sigma_2)\sin 2\theta \tag{7.14}$$

A careful study of equations (7.13) and (7.14) reveals that they can be represented by a circle of radius $(\sigma_1 - \sigma_2)/2$, if σ_θ is the horizontal axis and τ_θ is the vertical axis, as shown in Fig. 7.5. The centre of the circle is at a distance $(\sigma_1 + \sigma_2)/2$ from 0, as shown in Fig. 7.5.

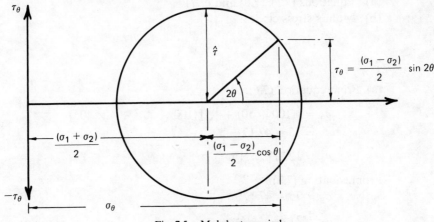

Fig. 7.5.   Mohr's stress circle.

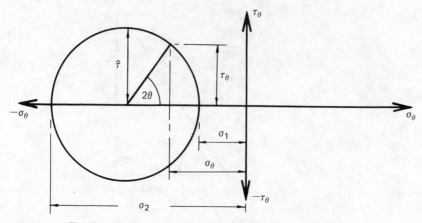

Fig. 7.6.   Mohr's stress circle for compressive principal stresses.

From Fig. 7.5, it can be seen that the maximum principal stress is on the right of the circle, and the minimum principal stress is on the left of the circle, but it must be pointed out that the magnitude of the minimum principal stress can be larger than the magnitude of the maximum principal stress, if the former is compressive, as shown in Fig. 7.6.

7.6.1   EXAMPLE 7.1   PRINCIPAL STRESSES

At a point in a piece of material, the state of stress is as follows:

$$\sigma_x = 100 \text{ MN/m}^2$$

$$\sigma_y = \ \ 50 \text{ MN/m}^2$$
$$\tau_{xy} = \ \ 40 \text{ MN/m}^2$$

Determine the direction and magnitudes of the principal stresses by

(a)   Equations (7.3), (7.6) and (7.7).
(b)   Mohr's stress circle.

7.6.2

(a) From equation (7.6):

$$\sigma_1 = \tfrac{1}{2}(100 + 50) + \tfrac{1}{2}\sqrt{[(100 - 50)^2 + 4 \times 40^2]}$$
$$= 75 + 47.17$$

$$\sigma_1 = 122.17 \text{ MN/m}^2$$

From equation (7.7):
$$\sigma_2 = 75 - 47.17$$

$$\underline{\sigma_2 = 27.83 \text{ MN/m}^2}$$

From equation (7.3):

$$\tan 2\theta = \frac{2 \times 40}{(100 - 50)} = 1.6$$

therefore

$$\theta = 28.997° = 29°$$

$$\hat{\tau} = (122.17 - 27.83)/2 = 47.17 \text{ MN/m}^2$$

7.6.3

(b) From a convenient point "0", determine the position of the point A, which has co-ordinates (σ_x, τ_{xy}), or (100, 40), as shown in Fig. 7.7. Next, determine the position of the point B, which has co-ordinates (σ_y, τ_{xy}), or (50, 40).

    The points A and B are on the circumference of Mohr's stress circle; so if they are joined together, they will form a chord on this circle.

    Bisect the line A–B and drop a perpendicular to obtain the centre "C" of Mohr's stress circle, as shown in Fig. 7.7. Using the point "C" as the centre of the circle, draw a circle through the points A and B, and hence measure off the values of σ_1, σ_2 and θ.

From Fig. 7.7, it can be seen that the values for σ_1, σ_2, θ and $\hat{\tau}$ are similar to those obtained in Section 7.6.2.

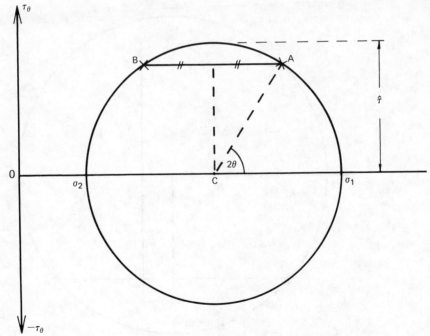

Fig. 7.7.

7.7.1   EXAMPLE 7.2   PRINCIPAL STRESSES VIA MOHR'S STRESS CIRCLE

At a point in a piece of material, the state of stress is as follows:

$$\sigma_x = 40 \text{ MN/m}^2$$
$$\sigma_y = -80 \text{ MN/m}^2$$
$$\tau_{xy} = -50 \text{ MN/m}^2$$

Determine σ_1, σ_2 and θ, from

(a) Equations (7.3), (7.6) and (7.7).
(b) Mohr's stress circle.

7.7.2

(a) From equation (7.6):

$$\sigma_1 = \tfrac{1}{2}(40 - 80) + \tfrac{1}{2}\sqrt{[(40 + 80)^2 + 4 \times 50^2]}$$
$$= -20 + 78.1$$
$$\underline{\sigma_1 = 58.1 \text{ MN/m}^2}$$

From equation (7.7):

$$\underline{\sigma_2 = -98.1 \text{ MN/m}^2}$$

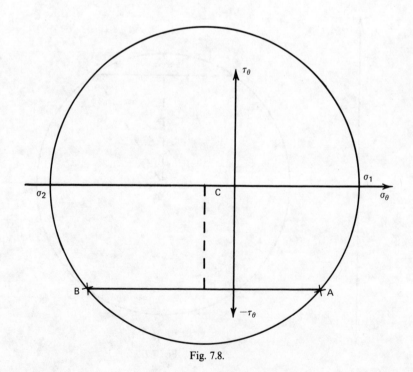

Fig. 7.8.

From equation (7.3):

$$\tan(2\theta) = -\tfrac{100}{120} = -0.8333$$

$$\theta = -19.9°$$

$$\hat{\tau} = (58.1 + 98.1)/2 = 78.1 \text{ MN/m}^2$$

7.7.3

(b) In a manner similar to that adopted in Example 7.1, the values of $\sigma_1, \sigma_2, \theta$ and $\hat{\tau}$ are obtained from Mohr's stress circle of Fig. 7.8.

From Fig. 7.8, it can be seen that the approximate values for $\sigma_1, \sigma_2, \theta$ and $\hat{\tau}$ are similar to those obtained for Section 7.7.2.

7.8.1   COMBINED BENDING AND TORSION

Problems in this category frequently occur in circular section shafts, particularly those of ships. In such cases, the combination of shearing stresses due to torsion and the direct stresses due to bending cause a complex system of stress.

Consider a circular section shaft, subjected to a combined bending moment M and a torque T, as shown in Fig. 7.9.

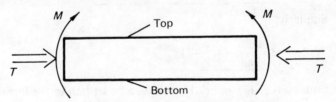

Fig. 7.9.   Shaft under combined bending and torsion.

The largest bending stresses due to M will be at both the top and the bottom of the shaft, and the combined effects of bending stress and shear stress at these positions will be as shown in Fig. 7.10.

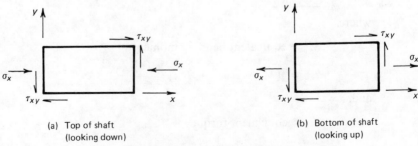

(a) Top of shaft
(looking down)

(b) Bottom of shaft
(looking up)

Fig. 7.10.   Complex stress system due to M and T.

Now, σ_x is entirely due to M, and τ_{xy} is entirely due to T, so that,

$$\sigma_x = M \cdot \left(\frac{64}{\pi d^4}\right) \cdot \frac{d}{2}$$

$$\sigma_x = \frac{32M}{\pi d^3} \tag{7.15}$$

and,

$$\tau_{xy} = T \cdot \left(\frac{32}{\pi d^3}\right) \cdot \frac{d}{2}$$

$$\tau_{xy} = \frac{16T}{\pi d^3} \tag{7.16}$$

From equilibrium considerations,

$$\sigma_y = 0$$

Substituting equations (7.15) and (7.16) into equations (7.6) and (7.7),

$$\sigma_1, \sigma_2 = \frac{16M}{\pi d^3} \pm \sqrt{\left[\left(\frac{16M}{\pi d^3}\right)^2 + \left(\frac{16T}{\pi d^3}\right)^2\right]}$$

$$\sigma_1, \sigma_2 = \frac{16}{\pi d^3}[M \pm \sqrt{(M^2 + T^2)}] \tag{7.17}$$

Similarly,

$$\hat{\tau} = \frac{16}{\pi d^3}(M^2 + T^2) \tag{7.18}$$

Equations (7.17) and (7.18) can also be written in the form:

$$\sigma_1, \sigma_2 = \frac{32 M_e}{\pi d^3} \tag{7.19}$$

and,

$$\hat{\tau} = \frac{16 T_e}{\pi d^3} \tag{7.20}$$

where,

$$M_e = \text{equivalent bending moment}$$

$$= \tfrac{1}{2}[M \pm \sqrt{(M^2 + T^2)}] \tag{7.21}$$

and,

$$T_e = \text{equivalent torque}$$

$$= \sqrt{(M^2 + T^2)} \tag{7.22}$$

7.9.1   EXAMPLE 7.3   SHIP'S PROPELLER SHAFT

A ship's propeller shaft is of a solid circular cross-section, of diameter 0.25 m. If the shaft is subjected to an axial thrust of 1 MN, together with a bending moment of 0.02 MN m and a torque of 0.05 MN m, determine the magnitude of the largest direct stress.

$$I = \frac{\pi \times 0.25^4}{64} = 1.917\text{E-4 m}^4$$

$$J = 3.83\text{E-4 m}^4$$

$$A = \frac{\pi \times 0.25^2}{4} = 0.0491 \text{ m}^2$$

The value of σ_x, due to bending, which is of interest, is the nagative value, because the axial thrust causes a compressive stress, and if these two are added together, the value of the largest σ_x will be given by

$$\sigma_x = -\frac{1}{0.0491} - \frac{0.02}{1.917\text{E-4}} \times 0.125$$

$$= -20.37 - 13.04$$

$$\sigma_x = -33.41 \text{ MN/m}^2 \tag{7.23}$$

$$\sigma_y = 0 \tag{7.24}$$

$$\tau_{xy} = \frac{0.05}{3.83\text{E-4}} \times 0.125 = \underline{16.32 \text{ MN/m}^2} \tag{7.25}$$

Substituting equations (7.23) to (7.25) into equations (7.6) and (7.7):

$$\sigma_1 = 6.65 \text{ MN/m}^2$$

$$\sigma_2 = -40.06 \text{ MN/m}^2$$

i.e. the largest magnitude of direct stress is compressive and equal to -40.06 MN/m^2.

7.10.1   EXAMPLE 7.4   COMPLEX STRESSES

The stresses σ_x, τ_{xy} and σ_1 are known on the triangular element "abc" of Fig. 7.11. Determine the principal stresses, and also σ_y and α.

As there is no shear stress on the plane "ab", σ_1 is a principal stress, and as σ_1 is greater than σ_x, σ_1 must be the maximum principal stress.

$$\sigma_1 = 200 \text{ MN/m}^2$$

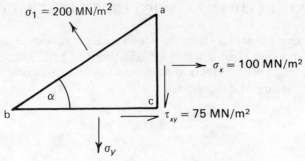

Fig. 7.11.

Resolving horizontally

$$200 * ab \sin \alpha = 75 * bc + 100 * ac$$

$$200 = 75 * \frac{bc}{ab \sin \alpha} + 100 * \frac{ac}{ab \sin \alpha}$$

$$= 75 \cot \alpha + 100$$

$$\cot \alpha = \tfrac{100}{75} \quad \text{or} \quad \tan \alpha = 0.75$$

therefore

$$\alpha = 36.87°$$

Resolving vertically

$$200 * ab \cos \alpha = 75 * ac + \sigma_y * bc$$

$$\sigma_y = 200 * \frac{ab}{bc} \cos \alpha - 75 * \frac{ac}{bc}$$

$$= 200 - 75 \tan \alpha$$

$$\sigma_y = 143.75 \text{ MN/m}^2.$$

From equation (7.7):

$$\sigma_2 = \frac{243.75}{2} - \tfrac{1}{2}\sqrt{(43.75^2 + 4 \times 75^2)}$$

$$= 121.88 - 78.13$$

$$\sigma_2 = 43.75 \text{ MN/m}^2$$

7.11.1 TWO-DIMENSIONAL STRAIN SYSTEMS

In a number of practical situations, particularly with experimental strain analysis, it is more convenient to make calculations with equations involving strains. Hence, for such cases, it will be necessary to obtain the expressions for strains, rather similar to those obtained for stresses in Section 7.3.2.

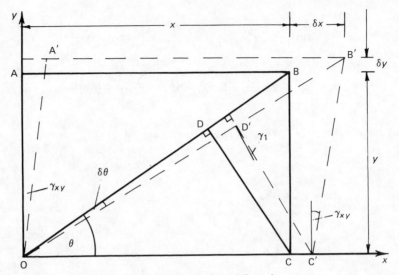

Fig. 7.12.   Strained quadrilateral.

Consider an infinitesimal rectangular elemental lamina of material OABC, in the x–y plane, which is subjected to an in-plane stress system, that causes the lamina to strain, as shown by the deformed quadrilateral OA′B′C′ of Fig. 7.12. Let the co-ordinate strains ε_x, ε_y and γ_{xy} be defined as follows:

$$\varepsilon_x = \text{direct strain in the } x \text{ direction}$$

$$\varepsilon_y = \text{direct strain in the } y \text{ direction}$$

$$\gamma_{xy} = \text{shear strain in the } x\text{–}y \text{ plane}$$

7.2.1   To Obtain ε_θ In Terms of the Co-ordinate Strains

Let

$$\varepsilon_\theta = \text{direct strain at angle } \theta$$
$$OB = r$$

so that,

$$BC = y = r \sin \theta$$
$$OC = x = r \cos \theta$$

Hence,

$$\delta_y = r \cdot \sin \theta \cdot \varepsilon_y$$

and,

$$\delta x = r \cdot \cos \theta \cdot \varepsilon_x + r \cdot \sin \theta \cdot \gamma_{xy}$$

7.12.2   Consider the Movement of B Parallel to OB

Now,

$$(OB)^2 = x^2 + y^2 = r^2$$

and,

$$(OB')^2 = (x + \delta x)^2 + (y + \delta y)^2 = (r + \delta r)^2$$

i.e.

$$x^2 + 2x \cdot \delta x + (\delta x)^2 + y^2 + 2y \cdot \delta y + (\delta y)^2 = r^2 + 2r \cdot \delta r + (\delta r)^2$$

or,

$$x^2 + y^2 + 2x \cdot \delta x + 2y \cdot \delta y = r^2 + 2r \cdot \delta r,$$

but as

$$x^2 + y^2 = r^2,$$

so that,

$$\delta r = \frac{x}{r} \cdot \delta x + \frac{y}{r} \cdot \delta y$$

or,

$$\delta r = \delta x \cos \theta + \delta y \sin \theta$$

Since the strains are small,

$$\varepsilon_\theta = \frac{\delta r}{r}$$

$$= \frac{\delta x}{r} \cdot \cos \theta + \frac{\delta y}{r} \sin \theta$$

Substituting for δx and δy,

$$\varepsilon_\theta = \varepsilon_x \cos^2 \theta + \varepsilon_y \sin^2 \theta + \gamma_{xy} \cdot \sin \theta \cos \theta$$

so that,

$$\underline{\varepsilon_\theta = \tfrac{1}{2}(\varepsilon_x + \varepsilon_y) + \tfrac{1}{2}(\varepsilon_x - \varepsilon_y) \cos 2\theta + \tfrac{1}{2} \gamma_{xy} \sin 2\theta} \qquad (7.26)$$

This is similar in form to the equation for stress at any angle θ.

7.12.3 To Determine the Shearing Strain γ_θ in Terms of Co-ordinate Strains

To evaluate shearing strain at an angle θ, we note that D is displaced to D'.
Now, as

$$BB' = \varepsilon_\theta \cdot r$$

and,

$$OD = OC \cos \theta = r \cos^2 \theta$$

then,

$$DD' = \varepsilon_\theta \cdot OD = \varepsilon_\theta \cdot r \cos^2 \theta$$

During straining, the line CD rotates anti-clockwise through a small angle γ_1, where,

$$\gamma_1 = \left(\frac{CC' \cos\theta - DD'}{CD}\right) = \left(\frac{\varepsilon_x \cos^2\theta - \varepsilon_\theta \cos^2\theta}{\cos\theta \sin\theta}\right)$$

Dividing the denominator into the numerator,

$$\gamma_1 = (\varepsilon_x - \varepsilon_\theta)\cot\theta$$

At the same time, OB rotates in a clockwise direction through a small angle, $\delta\theta$, as shown in Fig. 7.13.

$$\theta = \tan^{-1}\frac{y}{x}$$

$$\delta\theta = -\left(\frac{\partial\theta}{\partial x}\cdot\delta x + \frac{\partial\theta}{\partial y}\cdot\delta y\right)$$

Now,

$$\frac{\partial\theta}{\partial x} = \frac{-y/x^2}{1+\left(\dfrac{y}{x}\right)^2} = \frac{-y}{x^2+y^2} = \frac{-y}{r^2} = \frac{-\sin\theta}{r}$$

$$\frac{\partial\theta}{\partial y} = \frac{\dfrac{1}{x}}{1+\left(\dfrac{y}{x}\right)^2} = \frac{x}{x^2+y^2} = \frac{x}{r^2} = \frac{\cos\theta}{r}$$

therefore

$$\delta\theta = \frac{(\delta x \sin\theta - \delta y \cos\theta)}{r}$$

$$= \frac{[(r\cos\theta\cdot\varepsilon_x + \gamma_{xy}\cdot r\cdot\sin\theta)\sin\theta - (r\sin\theta\cdot\varepsilon_y)\cos\theta]}{r}$$

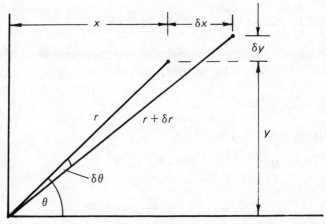

Fig. 7.13.

and shear strain at any angle θ,

$$\gamma_\theta = \gamma_1 + \delta\theta$$

$$\gamma_\theta = (\varepsilon_x - \varepsilon_\theta)\cot\theta + (\varepsilon_x \cos\theta + \gamma_{xy} \sin\theta)\sin\theta - \varepsilon_y \cdot \sin\theta \cos\theta$$

Substituting for ε_θ from equation (7.26), we get,

$$\gamma_\theta = \varepsilon_x \cot\theta - \tfrac{1}{2}(\varepsilon_x + \varepsilon_y)\cot\theta$$

$$- \tfrac{1}{2}(\varepsilon_x - \varepsilon_y)\cot\theta \cos 2\theta - \tfrac{1}{2}\gamma_{xy} \cdot \sin 2\theta \cdot \cot\theta$$

$$+ \varepsilon_x \cos\theta \sin\theta + \gamma_{xy} \sin^2\theta - \varepsilon_y \sin\theta \cos\theta$$

$$= \tfrac{1}{2}(\varepsilon_x - \varepsilon_y)\cot\theta - \tfrac{1}{2}(\varepsilon_x - \varepsilon_y)\cot\theta \cos 2\theta$$

$$- \sin\theta \cos\theta(-\varepsilon_x + \varepsilon_y) - \gamma_{xy}\left(\frac{\sin 2\theta \cdot \cot\theta}{2} - \sin^2\theta\right)$$

$$= \tfrac{1}{2}(\varepsilon_x - \varepsilon_y)\cot\theta\,(1 - \cos 2\theta) + (\varepsilon_x - \varepsilon_y)\frac{\sin 2\theta}{2}$$

$$- \gamma_{xy}\left(\sin\theta \cdot \cos\theta \cdot \frac{\cos\theta}{\sin\theta} - \sin^2\theta\right)$$

$$= (\varepsilon_x - \varepsilon_y)\cot\theta \cdot \sin^2\theta + (\varepsilon_x - \varepsilon_y)\frac{\sin 2\theta}{2} - \gamma_{xy}\cos 2\theta$$

$$= (\varepsilon_x - \varepsilon_y)\frac{\cot\theta}{\sin\theta} \cdot \sin^2\theta + \frac{(\varepsilon_x - \varepsilon_y)\sin 2\theta}{2} - \gamma_{xy}\cos 2\theta$$

$$\gamma_\theta = (\varepsilon_x - \varepsilon_y)\sin 2\theta - \gamma_{xy}\cos 2\theta \tag{7.27}$$

Equation (7.27) is often written in the form,

$$\frac{\gamma_\theta}{2} = \tfrac{1}{2}(\varepsilon_x - \varepsilon_y)\sin 2\theta - \tfrac{1}{2}\gamma_{xy}\cos 2\theta \tag{7.28}$$

7.13.1 PRINCIPAL STRAINS (ε_1 AND ε_2)

Principal strains are the maximum and minimum values of direct strain, and they are obtained by satisfying the equation:

$$\frac{d\varepsilon_\theta}{d\theta} = 0$$

Hence, from equation (7.26),

$$0 = -(\varepsilon_x - \varepsilon_y)\sin 2\theta + \gamma_{xy}\cos 2\theta$$

i.e.

$$\tan 2\theta = \frac{\gamma_{xy}}{(\varepsilon_x - \varepsilon_y)} \tag{7.29}$$

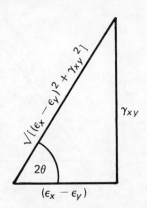

Fig. 7.14.   Mathematical triangle.

Equation (7.29) can be represented by the mathematical triangle of Fig. 7.14. From Fig. 7.14,

$$\sin 2\theta = \pm \frac{\gamma_{xy}}{\sqrt{[(\varepsilon_x - \varepsilon_y)^2 + \gamma_{xy}^2]}} \tag{7.30}$$

and

$$\cos 2\theta = \pm \frac{\varepsilon_x - \varepsilon_y}{\sqrt{[(\varepsilon_x - \varepsilon_y)^2 + \gamma_{xy}^2]}} \tag{7.31}$$

Substituting equations (7.30) and (7.31) into equation (7.28),

$$\gamma_\theta = 0$$

i.e. the shear strain on a principal plane is zero, as is the case for shear stress. Furthermore, as $\tau = G\gamma$, it follows that the *planes for principal strains are the same for principal stresses.*

7.13.2   To Obtain the Expressions for the Principal Strains in Terms of the Co-ordinate Strains

The values of the principal strains can be obtained by substituting equations (7.30) and (7.31) into equation (7.26):

$$\varepsilon_1, \varepsilon_2 = \tfrac{1}{2}(\varepsilon_x + \varepsilon_y) \pm \tfrac{1}{2}\sqrt{[(\varepsilon_x - \varepsilon_y)^2 + \gamma_{xy}^2]} \tag{7.32}$$

7.13.3   To Determine the Value and Direction of the Maximum Shear Strain ($\hat{\gamma}$)

The direction of $\hat{\gamma}$ can be obtained by satisfying the condition

$$\frac{\mathrm{d}\gamma_\theta}{\mathrm{d}\theta} = 0$$

Hence, from equation (7.28),

$$0 = (\varepsilon_x - \varepsilon_y)\cos 2\theta + \gamma_{xy} \sin 2\theta$$

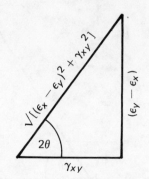

Fig. 7.15.   Mathematical triangle.

therefore

$$\tan 2\theta = -\frac{(\varepsilon_x - \varepsilon_y)}{\gamma_{xy}} \tag{7.33}$$

Equation (7.33) can be represented by the mathematical triangle of Fig. 7.15. From Fig. 7.15,

$$\sin 2\theta = \pm (\varepsilon_y - \varepsilon_x)/\sqrt{[(\varepsilon_x - \varepsilon_y)^2 + \gamma_{xy}^2]} \tag{7.34}$$

and,

$$\cos 2\theta = \pm \gamma_{xy}/\sqrt{[(\varepsilon_x - \varepsilon_y)^2 + \gamma_{xy}^2]} \tag{7.35}$$

Substituting equations (7.34) and (7.35) into equation (7.28),

$$\hat{\gamma} = \sqrt{[(\varepsilon_x - \varepsilon_y)^2 + \gamma_{xy}^2]} \tag{7.36}$$

which when compared with equation (7.32) reveals that,

$$\hat{\gamma} = \pm (\varepsilon_1 - \varepsilon_2) \tag{7.37}$$

If Fig. 7.15 is compared with Fig. 7.14, it can be seen that the maximum shear strain occurs at 45° to the principal planes.

7.14.1   MOHR'S CIRCLE OF STRAIN

On a principal plane,

$$\varepsilon_x = \varepsilon_1$$

$$\varepsilon_y = \varepsilon_2$$

and

$$\gamma_{xy} = 0$$

which when substituted into equations (7.26) and (7.28) yield the following expressions:

$$\varepsilon_\theta = \tfrac{1}{2}(\varepsilon_1 - \varepsilon_2) + \tfrac{1}{2}(\varepsilon_1 - \varepsilon_2)\cos 2\theta \tag{7.38}$$

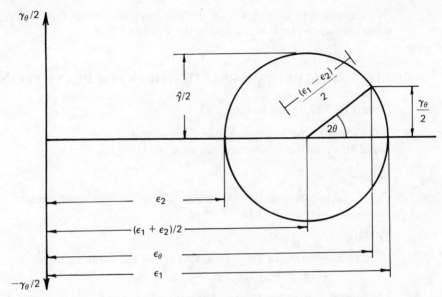

Fig. 7.16.   Mohr's circle of strain.

and,

$$\frac{\gamma_\theta}{2} = \frac{(\varepsilon_1 - \varepsilon_2)}{2} \sin 2\theta \tag{7.39}$$

Equations (7.38) and (7.39) can be represented by a circle if ε_θ is taken as the horizontal axis and $\gamma_\theta/2$ as the vertical axis, as shown in Fig. 7.16.

7.15.1

Plane stress is a two-dimensional system of stress and a three-dimensional system of strain, where the stresses σ_x and σ_y act in the plane of the plate, as shown in Fig. 7.17.

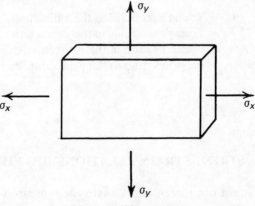

Fig. 7.17.   Plane stress.

In addition to causing strains in the x and y directions, these stresses will cause an out-of-plane strain due to the Poisson effect.

7.15.2 STRESS–STRAIN RELATIONSHIPS FOR PLANE STRESS

From Fig. 7.17, it can be seen that

the strain in the x direction due to $\sigma_x = \sigma_x/E$
and strain in the x direction due to $\sigma_y = -v\sigma_x/E$

so that,

$$\varepsilon_x = \text{strain in the } x \text{ direction, due to the combined stresses}$$
$$= (\sigma_x - v\sigma_y)/E \tag{7.40}$$

Similarly,

$$\varepsilon_y = \text{strain in the } y \text{ direction, due to the combined stresses}$$
$$= (\sigma_y - v\sigma_x)/E \tag{7.41}$$

The stress–strain relationships of equations (7.40) and (7.41) can be put in the alternative form of equations (7.42) and (7.43):

$$\sigma_x = \frac{E}{(1 - v^2)}(\varepsilon_x - v\varepsilon_y) \tag{7.42}$$

$$\sigma_y = \frac{E}{(1 - v^2)}(\varepsilon_y - v\varepsilon_x) \tag{7.43}$$

For an orthotropic material, equations (7.42) and (7.43) can be put in the form:

$$\sigma_x = \frac{1}{(1 - v_x v_y)}(E_x \varepsilon_x + v_x E_y \varepsilon_y) \tag{7.44}$$

$$\sigma_y = \frac{1}{(1 - v_x v_y)}(E_y \varepsilon_y + v_y E_x \varepsilon_x) \tag{7.45}$$

where,

$E_x = $ Young's modulus in the x direction
$E_y = $ Young's modulus in the y direction
$v_x = $ Poisson's ratio in the x direction, due to σ_y
$v_y = $ Poisson's ratio in the y direction, due to σ_x

and,

$$v_x E_y = v_y E_x \tag{7.46}$$

7.15.3 STRESS–STRAIN RELATIONSHIPS FOR PLANE STRAIN

Plane strain is a three-dimensional system of stress and a two-dimensional system of strain, as shown in Fig. 7.18, where the out-of-plane stress σ_z is

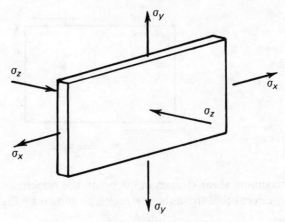

Fig. 7.18.   Plane strain.

related to the in-plane stresses σ_x and σ_y, and Poisson's ratio, so that the out-of-plane strain is zero. From Fig. 7.18,

$$\varepsilon_x = (\sigma_x - v\sigma_y - v\sigma_z)/E \tag{7.47}$$

$$\varepsilon_y = (\sigma_y - v\sigma_x - v\sigma_z)/E \tag{7.48}$$

$$\varepsilon_z = 0 = (\sigma_z - v\sigma_x - v\sigma_y)/E \tag{7.49}$$

From equation (7.49),

$$\sigma_z = v(\sigma_x + v\sigma_y) \tag{7.50}$$

Substituting equation (7.50) into equations (7.47) and (7.48), these last two equations can be put in the form of equations (7.51) and (7.52), which is usually found to be more convenient.

$$\sigma_x = \frac{E}{(1+v)(1-2v)}[(1-v)\varepsilon_x + v\varepsilon_y] \tag{7.51}$$

$$\sigma_y = \frac{E}{(1+v)(1-2v)}[(1-v)\varepsilon_y + v\varepsilon_x] \tag{7.52}$$

For both plane stress and plane strain,

$$\tau_{xy} = G\gamma_{xy} \tag{7.53}$$

7.16.1 PURE SHEAR

A system of pure shear is shown by Fig. 7.19, where the shear stresses τ are not accompanied by direct stresses on the same planes. As these shear stresses

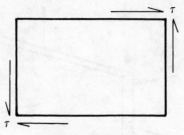

Fig. 7.19.   Pure shear.

are maximum shear stresses at the point, the principal stresses will lie at 45°
to the direction of these shear stresses, as shown by Fig. 7.20, where,

σ_1 = maximum principal stress
σ_2 = maximum principal stress

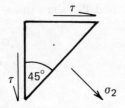 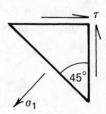

Fig. 7.20.

By resolution, it can be seen that,

$$\sigma_1 = \tau$$

and   (7.54)

$$\sigma_2 = -\tau$$

i.e. the system of shear stresses of Fig. 7.19 is equivalent to the system of Fig.
7.21. Thus, if shear strain is required to be measured, the strain guages have to
be placed at 45° to the directions of shear stresses, as shown in Fig. 7.22.
These strains will measure the principal strains, ε_1 and ε_2, and such strain
gauges are known as shear pairs.

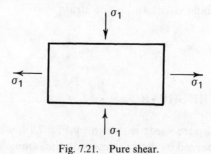

Fig. 7.21.   Pure shear.

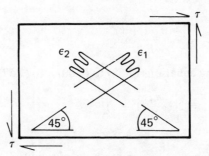

Fig. 7.22.　Pure shear.

From equation (7.37),

$$\gamma = \varepsilon_1 - \varepsilon_2 \tag{7.55}$$

and,

$$\tau = G\gamma \tag{7.56}$$

N.B.　If strain guages were placed in the directions of x or y, then the gauges would simply change their shapes, as shown in Fig. 7.23, and would not measure strain.

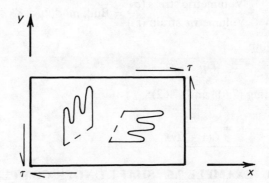

Fig. 7.23.　Incorrect method of shear strain measurement.

7.17.1　Relationships Between Elastic Constants

From equations (7.55) and (7.56):

$$\tau = G(\varepsilon_1 - \varepsilon_2) \tag{7.57}$$

but,

$$\sigma_1 = \tau \tag{7.58}$$

therefore

$$\varepsilon_1 = \frac{\sigma_1(1 + \nu)}{E} \tag{7.59}$$

and,

$$\varepsilon_2 = \frac{-\sigma_1(1 + v)}{E} \tag{7.60}$$

Substituting equations (7.58) to (7.60) into (7.57), the following is obtained:

$$\sigma_1 = 2G \cdot \frac{\sigma_1}{E}(1 + v)$$

therefore

$$G = \frac{E}{2(1 + v)} = \text{modulus of rigidity}$$

From Section 2.5.2,

$$\text{Volumetric strain} = \varepsilon_v = \varepsilon_x + \varepsilon_y + \varepsilon_z$$

But for an elemental cube under a hydrostatic stress σ,

$$\varepsilon_x = \varepsilon_y = \varepsilon_z = \sigma(1 - 2v)/E$$

therefore

$$\varepsilon_v = 3\sigma(1 - 2v)/E \tag{7.61}$$

Now,

$$\frac{\text{Volumetric stress }(\sigma)}{\text{Volumetric strain }(\varepsilon_v)} = \text{Bulk modulus }(K)$$

therefore

$$\varepsilon_v = \sigma/K \tag{7.62}$$

Equating (7.61) and (7.62):

$$K = \frac{E}{3(1 - 2v)} \tag{7.63}$$

7.18.1   EXAMPLE 7.5   SHAFT UNDER COMPLEX LOADING

A solid circular section rotating shaft, of diameter 0.2 m, is subjected to combined bending, torsion and axial load, where the maximum direct stresses due to bending occur on the top and bottom surfaces of the shaft.

If a shear pair is attached to the shaft, then the strains recorded from this pair are as follows:

$$\left.\begin{array}{l} \varepsilon_1^T = 200 \times 10^{-6} \\ \varepsilon_2^T = \ \ 80 \times 10^{-6} \end{array}\right\} \text{ when the shear pair is at the top}$$

$$\left.\begin{array}{l} \varepsilon_1^B = 100 \times 10^{-6} \\ \varepsilon_2^B = -20 \times 10^{-6} \end{array}\right\} \text{ when the shear pair is at the bottom}$$

$$E = 2 \times 10^{11} \text{ N/m}^2$$

$$v = 0.3$$

Let,

ε_{1T}^{T} and ε_{2T}^{T} = strain in gauges ① and ② due to T at the top

ε_{1T}^{B} and ε_{2T}^{B} = strain in gauges ① and ② due to T at the bottom

ε_{1M}^{T} and ε_{2M}^{T} = strain in gauges ① and ② due to M at the top

ε_{1M}^{B} and ε_{2M}^{B} = strain in gauges ① and ② due to M at the bottom

ε_{1D}^{T} and ε_{2D}^{T} = strain in gauges ① and ② due to the axial load at the top

ε_{1D}^{B} and ε_{2D}^{B} = strain in gauges ① and ② due to the axial load at the bottom

By inspection, it can be deduced that,

$$\varepsilon_{1T}^{T} = -\varepsilon_{2T}^{T}; \quad \varepsilon_{1T}^{B} = -\varepsilon_{2T}^{B}; \quad \varepsilon_{1T}^{T} = \varepsilon_{1T}^{B}; \qquad \varepsilon_{2T}^{T} = \varepsilon_{2T}^{B};$$

$$\varepsilon_{1M}^{T} = \varepsilon_{2M}^{T}; \quad \varepsilon_{1M}^{B} = \varepsilon_{2M}^{B}; \quad \varepsilon_{1M}^{T} = -\varepsilon_{1M}^{B}; \quad \varepsilon_{2M}^{T} = -\varepsilon_{2M}^{B};$$

$$\varepsilon_{1D}^{T} = \varepsilon_{2D}^{T} = \varepsilon_{1D}^{B} = \varepsilon_{2D}^{B}$$

Hence, *at the top*

$$\varepsilon_1^{T} = \varepsilon_{1T}^{T} + \varepsilon_{1M}^{T} + \varepsilon_{1D}^{T} \tag{7.64}$$

$$\varepsilon_2^{T} = -\varepsilon_{1T}^{T} + \varepsilon_{1M}^{T} + \varepsilon_{1D}^{T} \tag{7.65}$$

and *at the bottom,*

$$\varepsilon_1^{B} = \varepsilon_{1T}^{T} - \varepsilon_{1M}^{T} + \varepsilon_{1D}^{T} \tag{7.66}$$

$$\varepsilon_2^{B} = -\varepsilon_{1T}^{T} - \varepsilon_{1M}^{T} + \varepsilon_{1D}^{T} \tag{7.67}$$

7.18.2   To Obtain T

Taking equation (7.65) from equation (7.64), or equation (7.67) from equation (7.66):

$$\gamma = \varepsilon_1^{T} - \varepsilon_2^{T} = 200 \times 10^{-6} - 80 \times 100^{-6}$$

$$\underline{\gamma = 120 \times 10^{-6}}$$

Now,

$$G = \frac{E}{2(1 + v)} = \frac{2 \times 10^{11}}{2.6}$$

$$\underline{G = 7.69 \times 10^{10} \text{ N/m}^2}$$

therefore

$$\underline{\tau = G\gamma = 9.228 \text{ MN/m}^2}$$

Now,

$$J = \frac{\pi \times 0.2^4}{32} = 1.571 \times 10^{-4} \text{ m}^4$$

and,

$$T = \frac{\tau J}{r} = 14.497 \text{ kN m}$$

7.18.3 To Obtain M

Take equation (7.66) from equation (7.64), or equation (7.67) from equation (7.65), to give,

$$2\varepsilon_{1M}^{M} = \varepsilon_1^T - \varepsilon_1^B = \varepsilon_2^T - \varepsilon_2^B$$

or,

$$\varepsilon_{1M}^T = 50 \times 10^{-6}$$

Let,

σ_b' = stress due to bending, acting along the strain gauges, as shown in Fig. 7.24.

Consider the equilibrium of the triangle of Fig. 7.25.

$$\sigma_b' \times \sqrt{2} = \frac{\sigma_b}{2}$$

therefore

$$\sigma_b' = \sigma_b/2 \tag{7.68}$$

where,

σ_b = bending stress in x direction, due to M.

From equation (7.40):

$$\varepsilon_{1M}^M = \frac{1}{E}(\sigma_b' - v\sigma_b'$$

$$= \frac{\sigma_b'}{E}(1 - v) = \frac{\sigma_b(1 - v)}{2E} = 50 \times 10^{-6} \tag{7.69}$$

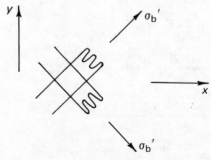

Fig. 7.24.

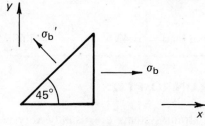

Fig. 7.25.

therefore

$$\sigma_b = \pm \frac{2 \times 2 \times 10^{11} \times 50 \times 10^{-6}}{0.7} = +28.57 \text{ MN/m}^2$$

Now,

$$M = \frac{\sigma}{y} * I$$

$$= \frac{28.57 * 7.855 \times 10^{-5}}{0.1}$$

$$M = 22.44 \text{ kN m}$$

7.18.4   To Obtain Direct Load

$$\varepsilon_1^T + \varepsilon_2^T = 2\varepsilon_{1M}^T + 2\varepsilon_1^T D \qquad (7.70)$$

and,

$$\varepsilon_1^B + \varepsilon_2^B = -2\varepsilon_{1M}^T + 2\varepsilon_1^T D \qquad (7.71)$$

Adding equations (7.70) and (7.71):

$$\varepsilon_1^T + \varepsilon_2^T + \varepsilon_1^B + \varepsilon_2^B = 4\varepsilon_1^T D$$

therefore

$$\varepsilon_{1D}^T = \frac{280 \times 10^{-6} + 80 \times 10^{-6}}{4} = 90 \times 10^{-6} \qquad (7.72)$$

In a manner similar to that adopted for the derivation of equations (7.68) and (7.69),

$$\varepsilon_{1D}^T = \frac{1}{2E} (\sigma_d - \nu\sigma_d) \qquad (7.73)$$

where,

$$\sigma_d = \text{stress due to axial load}$$

From equations (7.72) and (7.73),

$$\sigma_d = 51.43 \text{ MN/m}^2$$

therefore

$$\text{axial load} = 51.43 \times \frac{\pi \times 0.2^2}{4} = 1.62 \text{ MN (tensile)}$$

7.19.1 STRAIN ROSETTES

In practice, strain systems are usually very complicated, and when it is required to experimentally determine the stresses at various points in a plate, at least three strain gauges have to be used at each point of interest.

The reason for requiring at least three strain guages at each point in the plate is that there are three unknowns at each point, namely the two principal strains and their direction.

Thus, by inputting the values of the three strains, together with their "positions", into equation (7.26), three simultaneous equations will result, the solution of which will yield the two principal strains, ε_1 and ε_1, and their "direction" θ.

To illustrate the method of determining ε_1, ε_2 and θ, consider the strains of Fig. 7.26.

Let,

$$\left.\begin{array}{l} \varepsilon_1 = \text{maximum principal strain} \\ \varepsilon_2 = \text{minimum principal strain} \\ \theta = \text{angle between } \varepsilon_1 \text{ and } \varepsilon_\theta \end{array}\right\} \text{Unknowns to be determined}$$

$$\left.\begin{array}{l} \varepsilon_\theta = \text{direct strain at an angle } \theta \text{ from } \varepsilon_1 \\ \varepsilon_\alpha = \text{direct strain at an angle } \alpha \text{ from } \varepsilon_\theta \\ \varepsilon_\beta = \text{direct strain at an angle } \beta \text{ from } \varepsilon_\alpha \end{array}\right\} \begin{array}{l} \text{Experimentally} \\ \text{measured strains} \end{array} \quad (7.74)$$

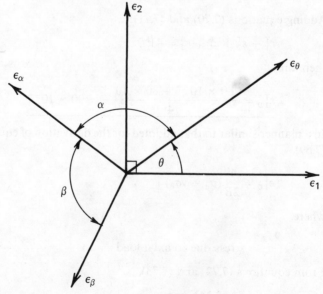

Fig. 7.26.

Substituting each measured value of equation (7.74), in turn, into equation (7.26), together with its "direction", the following three simultaneous equations are obtained:

$$\varepsilon_\theta = \tfrac{1}{2}(\varepsilon_1 + \varepsilon_2) + \tfrac{1}{2}(\varepsilon_1 - \varepsilon_2)\cos 2\theta \tag{7.75}$$

$$\varepsilon_\alpha = \tfrac{1}{2}(\varepsilon_1 + \varepsilon_2) + \tfrac{1}{2}(\varepsilon_1 - \varepsilon_2)\cos[2(\theta + \alpha)] \tag{7.76}$$

$$\varepsilon_\beta = \tfrac{1}{2}(\varepsilon_1 + \varepsilon_2) + \tfrac{1}{2}(\varepsilon_1 - \varepsilon_2)\cos[2(\theta + \alpha + \beta)] \tag{7.77}$$

From equations (7.75) to (7.77), it can be seen that the only unknowns are ε_1, ε_2 and θ, which can be readily determined.

For mathematical convenience, manufacturers of strain gauges normally supply strain rosettes, where the angles α and β have values of 45° or 60° or 120°. (The strain gauge technique is described in greater detail in Chapter 10.)

If, however, the experimentalist has difficulty in attaching a standard rosette to an awkward zone, then he/she can attach three linear strain gauges at this point, and choose suitable values for α and β. In such cases, it will be necessary to use equations (7.75) to (7.77) to determine the required unknowns.

7.19.2   45° Rectangular Rosette (See Fig. 7.27)

Substituting α and β into equations (7.75) to (7.77), the three simultaneous equations (7.78) to (7.80) are obtained:

$$\varepsilon_\theta = \tfrac{1}{2}(\varepsilon_1 + \varepsilon_2) + \tfrac{1}{2}(\varepsilon_1 - \varepsilon_2)\cos 2\theta \tag{7.78}$$

$$\varepsilon_\alpha = \tfrac{1}{2}(\varepsilon_1 + \varepsilon_2) - \tfrac{1}{2}(\varepsilon_1 - \varepsilon_2)\sin 2\theta \tag{7.79}$$

$$\varepsilon_\beta = \tfrac{1}{2}(\varepsilon_1 + \varepsilon_2) - \tfrac{1}{2}(\varepsilon_1 - \varepsilon_2)\cos 2\theta \tag{7.80}$$

Adding equation (7.78) to (7.80):
$$\varepsilon_\theta + \varepsilon_\beta = \varepsilon_1 + \varepsilon_2 \tag{7.81}$$

Equation (7.81) is known as the *first invariant of strain*, which states that the sum of mutually perpendicular direct strains at a point is constant.

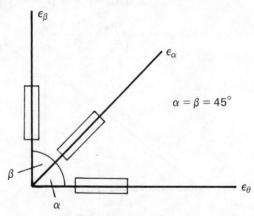

Fig. 7.27.   45° rectangular rosette.

From equations (7.79) and (7.80):

$$-\tfrac{1}{2}(\varepsilon_1 - \varepsilon_2)\sin 2\theta = \varepsilon_\alpha - \tfrac{1}{2}\varepsilon_1 - \tfrac{1}{2}\varepsilon_2 \tag{7.82}$$

$$-\tfrac{1}{2}(\varepsilon_1 - \varepsilon_2)\cos 2\theta \doteq \varepsilon_\beta - \tfrac{1}{2}\varepsilon_1 - \tfrac{1}{2}\varepsilon_2 \tag{7.83}$$

Substituting equation (7.81) into (7.82) and (7.83), and then dividing equation (7.82) by equation (7.83),

$$\tan 2\theta = \frac{\varepsilon_\alpha - \tfrac{1}{2}\varepsilon_\theta - \tfrac{1}{2}\varepsilon_\beta}{\varepsilon_\beta - \tfrac{1}{2}\varepsilon_\theta - \tfrac{1}{2}\varepsilon_\beta}$$

$$\tan 2\theta = \frac{(\varepsilon_\theta - 2\varepsilon_\alpha + \varepsilon_\beta)}{(\varepsilon_\theta - \varepsilon_\beta)} \tag{7.84}$$

7.19.3 To Determine ε_1 and ε_2

Taking equation (7.80) from equation (7.78),

$$\varepsilon_\theta - \varepsilon_\beta = (\varepsilon_1 - \varepsilon_2)\cos 2\theta \tag{7.85}$$

Hence, from equations (7.81) and (7.85):

$$\varepsilon_1 = \frac{\varepsilon_\theta + \varepsilon_\beta}{2} + \frac{\varepsilon_\theta - \varepsilon_\beta}{2\cos 2\theta} \tag{7.86}$$

$$\varepsilon_2 = \frac{\varepsilon_\theta + \varepsilon_\beta}{2} - \frac{\varepsilon_\theta - \varepsilon_\beta}{2\cos 2\theta} \tag{7.87}$$

Equation (7.84) can be represented by the mathematical triangle of Fig. 7.28, where,

$$\text{hypotenuse} = \sqrt{[\varepsilon_\theta^2 + 4\varepsilon_\alpha^2 + \varepsilon_\beta^2 - 4\varepsilon_\theta\varepsilon_\alpha - 4\varepsilon_\alpha\varepsilon_\beta}$$

$$+ 2\varepsilon_\theta\varepsilon_\beta + \varepsilon_\theta^2 + \varepsilon_\beta^2 - 2\varepsilon_\theta\varepsilon_\beta]$$

$$= \sqrt{2}\sqrt{[(\varepsilon_\theta - \varepsilon_\alpha)^2 + (\varepsilon_\beta - \varepsilon_\alpha)^2]} \tag{7.88}$$

$$\cos 2\theta = \frac{\varepsilon_\theta - \varepsilon_\beta}{\sqrt{2}\sqrt{[(\varepsilon_\theta - \varepsilon_\alpha)^2 + (\varepsilon_\beta - \varepsilon_\alpha)^2]}} \tag{7.89}$$

$$\sin 2\theta = \frac{\varepsilon_\theta - 2\varepsilon_\alpha + \varepsilon_\beta}{\sqrt{2}\sqrt{[(\varepsilon_\theta - \varepsilon_\alpha)^2 + (\varepsilon_\beta - \varepsilon_\alpha)^2]}} \tag{7.90}$$

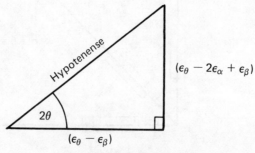

Fig. 7.28. Mathematical triangle.

Substituting equations (7.89) and (7.90) into equation (7.86) and (7.87):

$$\varepsilon_1, \varepsilon_2 = \tfrac{1}{2}(\varepsilon_\theta + \varepsilon_\beta) \pm \frac{\sqrt{2}}{2}\sqrt{[(\varepsilon_\theta - \varepsilon_\alpha)^2 + (\varepsilon_\beta - \varepsilon_\alpha)^2]} \tag{7.91}$$

7.19.4   120° Equiangular Rosette (See Fig. 7.29)

For greater precision, 120° equiangular rosettes are preferred to 45° rectangular rosettes.

Substituting α and β into equations (7.75) to (7.77), the simultaneous equations of (7.92) to (7.94) are obtained.

$$\varepsilon_\theta = \tfrac{1}{2}(\varepsilon_1 + \varepsilon_2) + \tfrac{1}{2}(\varepsilon_1 - \varepsilon_2)\cos 2\theta \tag{7.92}$$

$$\varepsilon_\alpha = \tfrac{1}{2}(\varepsilon_1 + \varepsilon_2) + \tfrac{1}{2}(\varepsilon_1 - \varepsilon_2)\cos[2(\theta + 120)] \tag{7.93}$$

$$\varepsilon_\beta = \tfrac{1}{2}(\varepsilon_1 + \varepsilon_2) + \tfrac{1}{2}(\varepsilon_1 - \varepsilon_2)\cos[2(\theta + 240)] \tag{7.94}$$

Equations (7.93) and (7.94) can be rewritten in the forms:

$$\varepsilon_\alpha = \tfrac{1}{2}(\varepsilon_1 + \varepsilon_2) + \tfrac{1}{2}(\varepsilon_1 - \varepsilon_2)(-\cos 2\theta \cdot \cos 60 + \sin 2\theta \cdot \sin 60) \tag{7.95}$$

$$\varepsilon_\beta = \tfrac{1}{2}(\varepsilon_1 + \varepsilon_2) + \tfrac{1}{2}(\varepsilon_1 - \varepsilon_2)(-\cos 2\theta \cdot \cos 60 - \sin 2\theta \cdot \sin 60) \tag{7.96}$$

Adding together equations (7.92), (7.95) and (7.96), we get,

$$\varepsilon_\theta + \varepsilon_\alpha + \varepsilon_\beta = \tfrac{3}{2}(\varepsilon_1 + \varepsilon_2)$$

$$\varepsilon_1 + \varepsilon_2 = \tfrac{2}{3}(\varepsilon_\theta + \varepsilon_\alpha + \varepsilon_\beta) \tag{7.97}$$

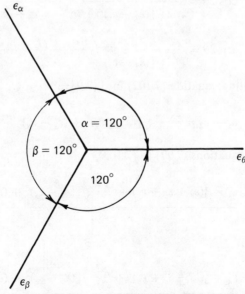

Fig. 7.29.   120° equiangular rosette.

Taking equation (7.96) from (7.95),

$$\varepsilon_\alpha - \varepsilon_\beta = (\varepsilon_1 + \varepsilon_2)\sin 2\theta \cdot \sin 60 \tag{7.98}$$

Taking equation (7.96) from (7.92),

$$\varepsilon_\theta - \varepsilon_\beta = \tfrac{1}{2}(\varepsilon_1 + \varepsilon_2)(\tfrac{3}{2}\cos 2\theta + \sin 2\theta \cdot \sin 60 \tag{7.99}$$

Dividing equation (7.99) by (7.98),

$$\frac{\varepsilon_\theta - \varepsilon_\beta}{\varepsilon_\alpha - \varepsilon_\beta} = \tfrac{1}{2}\left(\tfrac{3}{2} \cdot \frac{\cot 2\theta}{\sin 60} + 1\right)$$

$$\frac{2(\varepsilon_\theta - \varepsilon_\beta)}{\varepsilon_\alpha - \varepsilon_\beta} = \tfrac{3}{2} \cdot \frac{2}{\sqrt{3}}\cot (2\theta) + 1$$

therefore

$$\sqrt{3}\cot 2\theta = \frac{2(\varepsilon_\theta - \varepsilon_\beta)}{\varepsilon_\alpha - \varepsilon_\beta} - \frac{(\varepsilon_\alpha - \varepsilon_\beta)}{\varepsilon_\alpha - \varepsilon_\beta}$$

or,

$$\tan 2\theta = \frac{\sqrt{3}(\varepsilon_\alpha - \varepsilon_\beta)}{(2\varepsilon_\theta - \varepsilon_\beta - \varepsilon_\alpha)} \tag{7.100}$$

7.19.5 To Determine ε_1 and ε_2

Equation (7.100) can be represented by the mathematical triangle of Fig. 7.30.
From Fig. 7.30,

$$\cos 2\theta = \frac{2\varepsilon_\theta - \varepsilon_\alpha - \varepsilon_\beta}{\sqrt{2}\sqrt{[(\varepsilon_\theta - \varepsilon_\alpha)^2 + (\varepsilon_\alpha - \varepsilon_\beta)^2 + (\varepsilon_\theta - \varepsilon_\beta)^2]}} \tag{7.101}$$

$$\sin 2\theta = \frac{\sqrt{3}(\varepsilon_\alpha - \varepsilon_\beta)}{\sqrt{2}\sqrt{[(\varepsilon_\theta - \varepsilon_\alpha)^2 + (\varepsilon_\alpha - \varepsilon_\beta)^2 + (\varepsilon_\theta - \varepsilon_\alpha)^2]}} \tag{7.102}$$

Substituting equation (7.102) into (7.98),

$$\varepsilon_1 - \varepsilon_2 = \frac{2\sqrt{2}}{3}\sqrt{[(\varepsilon_\theta - \varepsilon_\alpha)^2 + (\varepsilon_\alpha - \varepsilon_\beta)^2 + (\varepsilon_\theta - \varepsilon_\beta)^2]} \tag{7.103}$$

Adding equations (7.97) to (7.103),

$$\varepsilon_1 = \tfrac{1}{3}(\varepsilon_\theta + \varepsilon_\alpha + \varepsilon_\beta) + \frac{\sqrt{2}}{3}\sqrt{[(\varepsilon_\theta - \varepsilon_\alpha)^2 + (\varepsilon_\alpha - \varepsilon_\beta)^2 + (\varepsilon_\theta - \varepsilon_\beta)^2]}$$

$$\tag{7.104}$$

Similarly,

$$\varepsilon_2 = \tfrac{1}{3}(\varepsilon_\theta + \varepsilon_\alpha + \varepsilon_\beta) - \frac{\sqrt{2}}{3}\sqrt{[(\varepsilon_\theta - \varepsilon_\alpha)^2 + (\varepsilon_\alpha - \varepsilon_\beta)^2 + (\varepsilon_\theta - \varepsilon_\beta)^2]}$$

$$\tag{7.105}$$

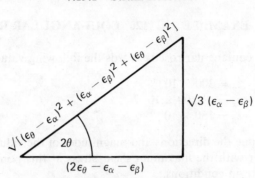

Fig. 7.30.   Mathematical triangle.

7.19.6   Other Types of Strain Gauge Rosette

These include the 60° delta of Fig. 7.31, the four-gauge 45° fan of Fig. 7.32 and the four-gauge T-delta of Fig. 7.33.

Advantages of using four-gauge rosettes are that they can lead to greater precision, particularly if one of the strains is zero.

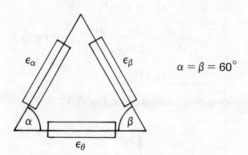

Fig. 7.31.   60° delta rosette.

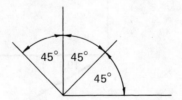

Fig. 7.32.   Four-gauge 45° fan rosette.

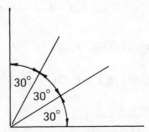

Fig. 7.33.   T-delta rosette.

7.20.1 EXAMPLE 7.6 120° EQUI-ANGULAR ROSETTE

A 120° equiangular rosette rcords the following values of strain at a point:

$$\varepsilon_\theta = 300 \times 10^{-6}$$
$$\varepsilon_\alpha = -100 \times 10^{-6}$$
$$\varepsilon_\beta = 150 \times 10^{-6}$$

Determine the directions and magnitudes of the values of principal stresses, together with the maximum shear stress at this point, for plane stress and plane strain conditions.

$$E = 2 \times 10^{11} \text{ N/m}^2 \quad v = 0.3$$

7.20.2 To Determine θ

From equation (7.100),

$$\tan 2\theta = \frac{\sqrt{3}(\varepsilon_\alpha - \varepsilon_\beta)}{(2\varepsilon_\theta - \varepsilon_\beta - \varepsilon_\alpha)}$$

$$= \frac{\sqrt{3} \times (-250)}{(600 - 150 + 100)} = -1.270$$

$$\underline{\theta = -19.11°}$$

For directions of σ_1 and σ_2, see Fig. 7.34.

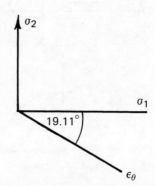

Fig. 7.34. Directions of σ_1 and σ_2.

From equation (7.104),

$$\varepsilon_1 = 10^{-6}\{116.7 + 0.471 \times 495\}$$

$$\underline{\varepsilon_1 = 350 \times 10^{-6}}$$

$$\underline{\varepsilon_2 = -116.4 \times 10^{-6}}$$

7.20.3   To Determine σ_1, σ_2 and $\hat{\tau}$ for Plane Stress

From equations (7.42) and (7.43),

$$\sigma_1 = \frac{2 \times 10^{11}}{0.91} (350 - 34.92) \times 10^{-6} = 69.25 \text{ MN/m}^2$$

$$\sigma_2 = \frac{2 \times 10^{11}}{0.91} (-116.4 + 105) \times 10^{-6} = -2.51 \text{ MN/m}^2$$

$$\hat{\tau} = \frac{\sigma_1 - \sigma_2}{2} = 35.88 \text{ MN/m}^2$$

7.20.4   To Determine σ_1, σ_2 and $\hat{\tau}$ for Plane Strain

From equations (7.51) and (7.52),

$$\sigma_1 = 80.8 \text{ MN/m}^2 \qquad \sigma_2 = 9.05 \text{ MN/m}^2$$

$$\hat{\tau} = 35.88 \text{ MN/m}^2$$

7.21.1   Computer Program for Principal Stresses and Strains

Table 7.1 gives a listing of a computer program in BASIC, for determining principal stresses and strains from co-ordinate stresses or co-ordinate strains.

7.21.2

The *input* for determining principal stresses from co-ordinate stresses is as follows:

> Type in σ_x (direct stress in the x direction)
> Type in σ_y (direct stress in the y direction)
> Type in τ_{xy} (shear stress in the x–y plane)

7.21.3

Typical input and output values are given in Tables 7.2 and 7.3 for Examples 7.1 and 7.2.

7.21.4

The input for determining principal stresses and strains from co-ordinate strains is as follows:

> Type in ε_x (direct strain in x direction)
> Type in ε_y (direct strain in y direction)
> Type in γ_{xy} (shear strain in the x–y plane)
> Type in E (elastic modulus)
> Type in v (Poisson's ratio)

N.B.   When this part of the program is being used, the condition of *plane stress* is assumed.

Table 7.1. Computer program for calculating σ_1, σ_2, etc. from either the co-ordinate stresses or the co-ordinate strains.

```
100 REMark principal stresses & strains
110 CLS
120 PRINT:PRINT"principal stresses & strains"
130 PRINT:PRINT"copyright of C.T.F.Ross":PRINT
140 PRINT"if inputting CO-ORDINATE STRESSES type 1; else if inputting CO-ORDINAT
E STRAINS, type 0":PRINT
150 INPUT str
160 IF str=1 OR str=0 THEN GO TO 180
170 PRINT:PRINT"incorrect data":PRINT:GO TO 140
180 IF str=0 THEN GO TO 300
190 PRINT:PRINT"stress in x direction=";:INPUT sigmax
200 PRINT"stress in y direction=";:INPUT sigmay
210 PRINT"shear stress in x-y plane=";:INPUT tauxy
220 const=.5*SQRT((sigmax-sigmay)^2+4*tauxy^2)
230 sigma1=.5*(sigmax+sigmay)+const
240 sigma2=.5*(sigmax+sigmay)-const
250 theta=.5*ATAN(2*tauxy/(sigmax-sigmay))
260 GO TO 600
300 PRINT:PRINT"strain in x direction=";:INPUT ex
310 PRINT"strain in y direction=";:INPUT ey
320 PRINT"shear in x-y plane=";:INPUT gxy
330 const=.5*SQRT((ex-ey)^2+gxy^2)
340 e1=.5*(ex+ey)+const
350 e2=.5*(ex+ey)-const
360 theta=.5*ATAN(gxy/(ex-ey))
370 PRINT:PRINT"elastic modulus=";:INPUT e
380 PRINT"poisson's ratio=";:INPUT nu
600 OPEN£3,ser1
610 IF str=1 THEN PRINT£3:PRINT£3:PRINT£3,"principal stresses":PRINT£3
620 IF str=0 THEN PRINT£3:PRINT£3:PRINT£3,"principal stresses & principal strain
s":PRINT£3:GO TO 800
630 PRINT£3,"stress in x direction=";sigmax
640 PRINT£3,"stress in y direction=";sigmay
650 PRINT£3,"shear in x-y plane=";tauxy
660 PRINT£3:PRINT£3
670 PRINT£3,"maximum principal stress=";sigma1
680 PRINT£3,"minimum principal stress=";sigma2
690 PRINT£3,"theta=";theta*180/PI;" degrees"
695 PRINT£3,"maximum shear stress="; (sigma1-sigma2)/2
700 GO TO 1000
800 PRINT£3,"strain in x direction=";ex
810 PRINT£3,"strain in y direction=";ey
820 PRINT£3,"shear strain in x-y plane=";gxy
830 PRINT£3,"elastic modulus=";e
840 PRINT£3,"poisson's ratio=";nu
850 PRINT£3:PRINT£3
860 PRINT£3,"maximum principal strain=";e1
870 PRINT£3,"minimum principal strain=";e2
875 PRINT£3,"theta=";theta*180/PI;" degrees"
880 PRINT£3,"maximum shear strain=";e1-e2
890 sigma1=e*(e1+nu*e2)/(1-nu^2)
900 sigma2=e*(e2+nu*e1)/(1-nu^2)
910 PRINT£3,"maximum principal stress=";sigma1
920 PRINT£3,"minimum principal stress=";sigma2
930 PRINT£3,"maximum shear stress="; (sigma1-sigma2)/2
1000 PRINT£3:PRINT£3:PRINT£3:PRINT£3
1010 CLOSE£3
1020 STOP
```

Table 7.2.   Computer output for example 7.1.

```
principal stresses

stress in x direction=100
stress in y direction=50
shear in x-y plane=40


maximum principal stress=122.1699
minimum principal stress=27.83009
theta=28.99731 degrees
maximum shear stress=47.16991
```

Table 7.3.   Computer output for example 7.2.

```
principal stresses

stress in x direction=40
stress in y direction=-80
shear in x-y plane=-50


maximum principal stress=58.1025
minimum principal stress=-98.1025
theta=-19.90279 degrees
maximum shear stress=78.1025
```

EXAMPLES FOR PRACTICE 7

1. Prove that the following relationships apply to the 60° delta rosette of Fig. 31.

$$\tan 2\theta = \frac{\sqrt{3}(\varepsilon_\beta - \varepsilon_\alpha)}{(2\varepsilon_\theta - \varepsilon_\alpha - \varepsilon_\beta)}$$

$$\varepsilon_1, \varepsilon_2 = \tfrac{1}{3}(\varepsilon_\theta + \varepsilon_\alpha + \varepsilon_\beta) \pm \frac{\sqrt{2}}{3}\sqrt{[(\varepsilon_\theta - \varepsilon_\alpha)^2 + (\varepsilon_\alpha - \varepsilon_\beta)^2}$$

$$+ (\varepsilon_\theta - \varepsilon_\beta)^2]$$

2. At a certain point "A" in a piece of material, the magnitudes of the direct stresses are -10 MN/m^2, 30 MN/m^2 and 40 MN/m^2, as shown in Fig. Q.7.2.

    Determine the magnitude and direction of the principal stresses and the maximum shear stress.

{Portsmouth Polytechnic, March 1982}

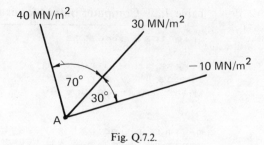

Fig. Q.7.2.

$$\{\sigma_1 = 62.67 \text{ MN/m}^2, \sigma_2 = -21.11 \text{ MN/m}^2, \hat{\tau} = 41.89 \text{ MN/m}^2\}$$

3. The web of a rolled steel joist has three linear strain gauges attached to the point "A" and indicating strains as shown in Fig. Q.7.3.

   Determine the magnitude and direction of the principal stresses, and the value of the maximum shear stress at this point, assuming the following to apply:

   Elastic modulus $= 2 \times 10^{11} \text{ N/m}^2$
   Poisson's ratio $= 0.3$

{Portsmouth Polytechnic, 1982}

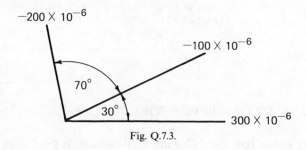

Fig. Q.7.3.

$$\{\sigma_1 = 62.2 \text{ MN/m}^2, \sigma_2 = -66.7 \text{ MN/m}^2, -21.35°, \hat{\tau} = 64.45 \text{ MN/m}^2\}$$

4. A solid stainless steel propeller shaft of a power boat is of diameter 2 cm. A 45° rectangular rosette is attached to the shaft, where the central gauge (Gauge No. 2) is parallel to the axis of the shaft.

   Assuming bending and thermal stresses are negligible, determine the thrust and torque that the shaft is subjected to, given that the recorded strains are as follows:

   $$\varepsilon_1 = -300 \times 10^{-6}$$
   $$\varepsilon_2 = -142.9 \times 10^{-6} \text{ (Gauge No. 2)}$$
   $$\varepsilon_3 = 200 \times 10^{-6}$$
   $$E = 2 \times 10^{11} \text{ N/m}^2$$
   $$v = 0.3$$

$$\{-8.98 \text{ kN}, 60.4 \text{ MN m}\}$$

5. At a point in a two-dimensional stress system, the known stresses are shown in Fig. Q.7.5.

Determine the principal stresses σ_y, α and $\hat{\tau}$.

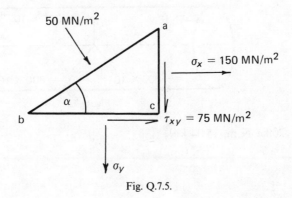

Fig. Q.7.5.

$$\{-20.56°, \sigma_1 = 178 \text{ MN/m}^2, \sigma_2 = -50 \text{ MN/m}^2, \sigma_y = 21.9 \text{ MN/m}^2\}$$

6. A solid circular section steel shaft, of 0.03 m diameter, is simply-supported at its ends and is subjected to a torque and a load that is radial to its axis. A small 60° strain gauge rosette is attached to the underside of the shaft at mid-span, as shown in Fig. Q.7.6.

Determine:

(a) the length of the shaft;
(b) the direction and value of the maximum principal stress;
(c) the applied torque.

$$E = 2.1 \times 10^{11} \text{ N/m}^2 \quad v = 0.28 \quad \rho = 7860 \text{ kg/m}^3,$$
$$g = 9.81 \text{ m/s}^2$$

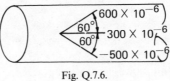

Fig. Q.7.6.

$\{5.08 \text{ m}, 37.65°$ anti-clockwise from the middle gauge, 146.5 MN/m^2, $-68.9 \text{ MN/m}^2, 552.3 \text{ N m}\}$

7. A solid cylindrical aluminium alloy shaft of 0.03 m diameter is subjected to a combined radial and axial load and a torque.

A 120° strain gauge rosette is attached to the shaft and records the strains shown in Fig. Q.7.7.

Determine the values of the axial load and torque, given the following:

$$E = 1 \times 10^{11} \text{ N/m}^2 \quad v = 0.32$$

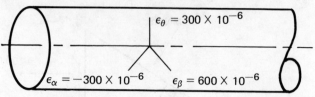

Fig. Q.7.7.

{208.7 N m, 15.04 kN}

8

Membrane Theory for Thin-walled Circular Cylinders and Spheres

8.1.1

The use of thin-walled circular cylinders and spheres for containing gases or liquids under pressure is a popular industrial requirement. This is partly because of the nature of fluid pressure and partly because such loads can be most efficiently resisted by in-plane membrane stresses, acting in curved shells.

In general, thin-walled shells have a very small bending resistance in comparison with their ability to resist loads in membrane tension. Although thin-walled shells also have a relatively large capability to resist pressure loads in membrane compression, the possibility arises that in such cases, failure can take place owing to structural instability (buckling) at stresses which may be a small fraction of that to cause yield, as shown in Fig. 8.1.

8.1.2

Thin-walled circular cylinders and spheres appear in many forms, including submarine pressure hulls, the legs of off-shore drilling rigs, containment vessels for nuclear reactors, boilers, condensers, storage tanks, gas holders, pipes, pumps and many other different types of pressure vessel.

Ideally, from a structural viewpoint, the perfect vessel to withstand uniform internal pressure is a thin-walled spherical shell, but such a shape may not necessarily be the most suitable from other considerations. For example, a submarine pressure hull, in the form of a spherical shell, is not a suitable shape for hydrodynamic purposes, nor for containing large quantities of equipment nor large numbers of personnel, and in any case, docking a spherically shaped vessel may present problems.

Fig. 8.1. Buckled forms of thin-walled cylinders under uniform external pressure.

Furthermore, pressure vessels of spherical shape may present difficulties in housing or storage, or in transport, particularly if the pressure vessel is being carried on the back of a lorry ("truck" in the U.S.A.).

Another consideration in deciding whether the pressure vessel should be cylindrical or spherical is from the point of view of its cost of manufacture. For example, although a spherical pressure vesel may be more structurally efficient than a similar cylindrical pressure vessel, the manufacture of the former may be considerably more difficult than the latter, so that additional labour costs of constructing a spherical pressure vessel may be much greater than any material savings that may be gained, especially as extruded cylindrical tubes can often be purchased "off the shelf".

8.1.3

Circular cylindrical shells are usually blocked off by domes, but can be blocked off by circular plates; however, if circular plates are used, their thickness is relatively large in comparison to the thickness of shell dome ends.

8.2.1 CIRCULAR CYLINDRICAL SHELLS UNDER UNIFORM   INTERNAL PRESSURE

The two major methods of failure of a circular cylindrical shell under uniform pressure are as follows.

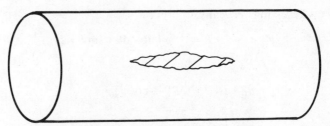

Fig. 8.2.   Fracture due to hoop stress.

8.2.2 (a)   Failure Due to Circumferential or Hoop Stress (σ_H)

If failure is due to the hoop stress, then fracture occurs along a longitudinal seam, as shown in Fig. 8.2.

Consider the circular cylinder of Fig. 8.3, which may be assumed to split in half, along two longitudinal seams, owing to the hoop stress, σ_H.

To determine σ_H, in terms of the applied pressure and the geometrical properties of the cylinder, consider the equilibrium of one half of the cylindrical shell, as shown in Fig. 8.4.

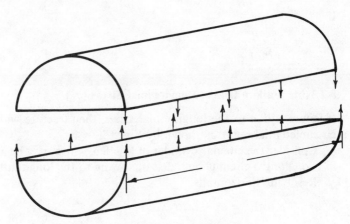

Fig. 8.3.   Failure due to hoop stress

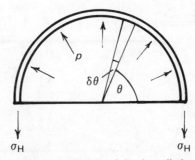

Fig. 8.4.   Equilibrium of circular cylinder.

Resolving vertically,

$$\sigma_H * t * 2 * L = \int_0^\pi p * R * d\theta * \sin\theta * L$$

or,

$$\sigma_H * t * 2 = pR[-\cos\theta]_0^\pi$$

$$= 2pR$$

therefore

$$\sigma_H = \frac{pR}{t} \tag{8.1}$$

where

σ_H = hoop stress, which under internal pressure is a maximum principal stress

p = internal pressure

R = internal radius

t = wall thickness of cylinder

L = length of cylinder

η_L = the structural efficiency of a longitudinal joint of the cylinder ($\eta_L \leqslant 1$), then

If

$$\underline{\underline{\sigma_H = \frac{pR}{\eta_L t}}} \tag{8.2}$$

8.2.3(b) Failure Due to Longitudinal Stress (σ_L)

If failure is due to the longitudinal stress, then fracture will occur along a circumferential seam, as shown by Fig. 8.5

Consider the circular cylinder of Fig. 8.6, which may be assumed to split in two, along a circumferential seam, owing to the longitudinal stress, σ_L.

Resolving horizontally,

$$p * \pi R^2 = \sigma_L * 2\pi Rt$$

therefore

$$\underline{\underline{\sigma_L = \frac{pR}{2t}}} \tag{8.3}$$

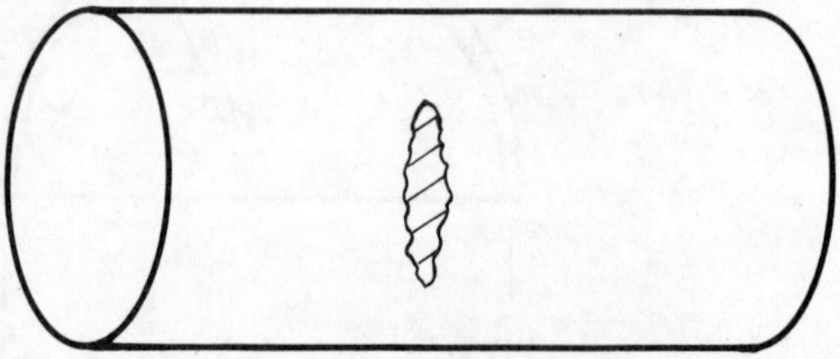

Fig. 8.5. Fracture due to longitudinal stress.

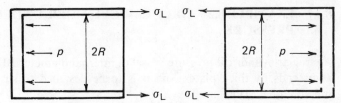

Fig. 8.6.   Failure along a circumferential seam.

where,

$$\sigma_L = \text{longitudinal stress}$$
$$= \text{minimum principal stress, if the pressure is internal}$$

Comparing equations (8.1) and (8.3), it can be seen that the *hoop stress has twice the magnitude of the longitudinal stress.*

If,

η_c = the structural efficiency of a circumferential joint of the cylinder ($\eta_c \leqslant 1$), then

$$\sigma_L = \frac{pR}{2\eta_c t} \tag{8.4}$$

8.3.1   EXAMPLE 8.1   EXTRUDED TUBE UNDER INTERNAL PRESSURE

Two identical extruded tubes of internal radius 5 m and of wall thickness 2 cm are joined together along a circumferential seam to form the main body of a pressure vessel. Assuming that the ends of the vessel are blocked off by very thick inextensible plates, determine the maximum stress in the vessel, when it is subjected to an internal pressure of 0.4 MPa.

It may be assumed that the joint efficiency equals 52%.

Now as the tubes are extruded, there is no longitudinal joint, so that,

$$\eta_L = 1$$

Therefore, from equation (8.2):

$$\sigma_H = \text{hoop stress} = \frac{0.4 \times 10^6 \times 5}{2 \times 10^{-2}}$$

$$\sigma_H = 100 \text{ MN/m}^2$$

Now,

$$\eta_c = 0.52$$

Therefore from equation (8.4):

$$\sigma_L = \text{longitudinal stress} = \frac{0.4 \times 10^6 \times 5}{2 \times 0.52 \times 2 \times 10^{-2}}$$

$$\sigma_L = 96.2 \text{ MN/m}^2$$

i.e. Maximum stress = $\sigma_H = 100$ MN/m^2

8.4.1   EXAMPLE 8.2   CIRCULAR CYLINDER UNDER INTERNAL PRESSURE

A circular cylindrical pressure vessel of internal diameter 10 m is blocked off at its ends by thick plates, and it is to be designed to sustain an internal pressure of 1 MPa.

Assuming the following to apply, determine a suitable value for the wall thickness of the cylinder:

(a) Longitudinal joint efficiency   =   98%
(b) Circumferential joint efficiency =   46%
(c) Maximum permissible stress   = 100 MN/m^2

8.4.2   Consideration of Hoop Stress

From equation (8.2):

$$\sigma_H = \frac{pR}{\eta_L t}$$

therefore

$$t = \frac{pR}{\eta_L \sigma_H} = \frac{1 \times 5}{0.98 \times 100}$$

$$\underline{t = 0.051 \text{ m}}$$

8.4.3   Consideration of Longitudinal Stress

From equation (8.4):

$$\sigma_L = \frac{pR}{2\eta_c t}$$

or,

$$t = \frac{pR}{2\eta_c \sigma_L}$$

$$= \frac{1 \times 5}{1 \times 0.46 \times 100}$$

$$\underline{t = 0.0544 \text{ m}}$$

i.e. $\underline{\underline{\text{Design wall thickness} = 5.44 \text{ cm}}}$

8.5.1   THIN-WALLED SPHERICAL SHELLS UNDER UNIFORM INTERNAL PRESSURE

Under uniform internal pressure, a thin-walled spherical shell of constant thickness will have a constant membrane tensile stress, where all such stresses will be principal stresses.

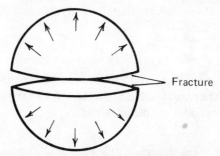

Fig. 8.7.   Fracture of spherical shell.

Let,

σ = membrane stress in the spherical shell

For such structures, fracture will occur along a diameter, as shown in Fig. 8.7.

8.5.2

To determine the membrane principal stress σ, in terms of the applied pressure p, and the geometry of the spherical shell, consider the equilibrium of one half of the spherical shell, as shown in Fig. 8.8.
Resolving vertically

$$\sigma * 2\pi Rt = \int_0^{\pi/2} p * 2\pi b * R \cdot d\theta * \sin\theta$$

but,

$$b = R\cos\theta$$

therefore

$$\sigma * 2\pi Rt = 2\pi R^2 p \int_0^{\pi/2} \cos\theta \sin\theta \, d\theta$$

or,

$$\sigma t = -pR \int_0^{\pi/2} \cos\theta \, d(\cos\theta)$$

$$= pR\left[-\frac{\cos^2\theta}{2}\right]_0^{\pi/2}$$

$$= \frac{pR}{2}$$

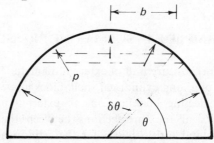

Fig. 8.8.   Equilibrium of hemispherical shell

therefore

$$\sigma = \frac{pR}{2t} \tag{8.5}$$

From equations (8.1) and (8.5), it can be seen that the maximum principal stress in a spherical shell has half the value of the maximum principal stress in a circular cylinder of the same radius.

If,

η = the structural efficiency of a joint on a diameter of the spherical shell ($\eta \leqslant 1$), then,

$$\sigma = \frac{pR}{2\eta t} \tag{8.6}$$

8.6.1   EXAMPLE 8.3   SPHERICAL SHELL UNDER INTERNAL PRESSURE

A thin-walled spherical vessel, of internal diameter 10 m, is to be designed to withstand an internal pressure of 1 MN/m^2.

Assuming the following apply, determine a suitable value of the wall thickness:

(a) Joint efficiency   = 75%
(b) Maximum permissible stress = 100 MN/m^2

From equation (8.6):

$$\sigma = \frac{pR}{2\eta t}$$

or,

$$t = \frac{pR}{2\eta\sigma} = \frac{1 \times 5}{2 \times 0.75 \times 100} = 0.033 \text{ m}$$

$$\underline{t = 3.3 \text{ cm}}$$

8.7.1   EXAMPLE 8.4   SUBMARINE PRESSURE HULL

A submarine pressure hull of external diameter 10 m may be assumed to be composed of a long cylindrical shell, blocked off by two hemispherical shell domes. Neglecting buckling due to external pressure, and the effects of discontinuity at the intersections between the domes and the cylinder, determine suitable thicknesses for the cylindrical shell body and the hemispherical dome ends.

The following may be assumed:

Maximum permissible stress   = 200 MN/m²
Diving depth of submarine   = 250 m
Density of sea water   = 1020 kg/m³
Acceleration due to gravity g = 9.81 m/s²
Longitudinal joint efficiency   = 90%
Circumferential joint efficiency = 70%

8.7.2   Cylindrical Body

From equation (8.2):

$$t = \frac{pR}{\eta_{L} * \sigma_{H}}$$

but,

$$p = 250 \text{ m} \times 1020 \frac{\text{kg}}{\text{m}^3} \times 9.81 \frac{\text{m}}{\text{s}^2}$$

$$= 2.5 \text{ MPa}$$

therefore

$$t = \frac{2.5 \times 10^6 \times 5}{0.9 \times 200 \times 10^6}$$

$$\underline{t = 0.0694 \text{ m} = 6.94 \text{ cm}}$$

8.7.3   Dome Ends

From equation (8.6):

$$t = \frac{pR}{2\eta\sigma}$$

$$= \frac{2.5 \times 10^6 \times 5}{2 \times 0.7 \times 200 \times 10^6}$$

$$\underline{t = 0.0446 \text{ m} = 4.46 \text{ cm}}$$

i.e.   Wall thickness of cylindrical body = 6.94 cm
Wall thickness of dome ends   = 4.46 cm

N.B.   From the above calculations, it can be seen that the required wall thickness of a submarine pressure hull increases roughly in proportion to its diving depth; thus, to ensure that the submarine has a sufficient reserve buoyancy for a given diameter, it is necessary to restrict its diving depth or to

Fig. 8.9. Axisymmetric buckling of an oblate dome under uniform external pressure.

use a material of construction which has a better strength : weight ratio. It should also be noted that under external pressure, the cylindrical section of a submarine pressure hull can buckle at a pressure which may be a fraction of that to cause yield, as shown in Fig. 8.1. In a similar manner, the dome ends can also buckle at a pressure which may be a small fraction of that to cause yield, as shown in Figs 8.9 and 8.10.

Fig. 8.10. Lobar buckling of a hemispherical or prolate dome under uniform external pressure.

8.8.1   Liquid Required to Raise the Pressure Inside a Circular Cylinder (Based on Small Deflection Elastic Theory)

In the case of a circular cylinder, assuming that it is just filled with liquid, the additional liquid that will be required to raise the internal pressure will be partly as a result of the swelling of the structure and partly as a result of the compression of the liquid itself, as shown by the following components:

(a) Longitudinal extension of the cylinder.
(b) Radial extension of the cylinder.
(c) Compressibility of the liquid.

Consider a thin-walled circular cylinder, blocked off at its ends by thick inextensible plates, and just filled with the liquid.

The calculation for the additional liquid to raise the internal pressure to p will be composed of the following three components.

8.8.2   Longitudinal Extension of the Circular Cylinder

Let,

u = The longitudinal movement of one end of the cylinder, relative to the other, as shown in Fig. 8.11.

ε_L = longitudinal strain in cylinder

$= u/L$ \hfill (8.7)

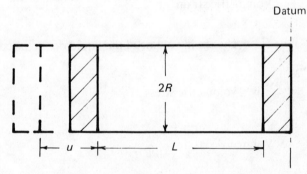

Fig. 8.11.   Longitudinal movement of the cylinder.

Change of volume of cylinder, due to u =

$$\delta V^1 = \pi R^2 u = \pi R^2 \varepsilon_L \cdot L \hfill (8.8)$$

8.8.3   Radio Extension of Cylinder

Let,

w = radial deflection of the cylinder, as shown in Fig. 8.12.

$$\varepsilon_H = \text{hoop strain} = \frac{2\pi(R + w) - 2\pi R}{2\pi R}$$

$$= \frac{w}{R} \hfill (8.9)$$

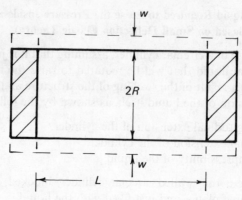

Fig. 8.12.   Radial deflection of the cylinder.

Change of volume of cylinder, due to w =

$$\delta V_2 = 2\pi R\,Lw = 2\pi R^2\,L\varepsilon_{\mathrm{H}}$$ (8.10)

8.8.4   Compressibility of Liquid

The liquid will compress owing to the raising of its pressure by p.
   From Section 2.5.1,

$$\frac{\text{Volumetric stress}}{\text{Volumetric strain}} = K = \text{Bulk modulus}$$

Let,

   δV_3 = change in volume of the liquid due to its compressibility

Now,

   p = volumetric stress

and,

   Volumetric strain = $\dfrac{\delta V_3}{V}$.

where,

   V = internal volume of cylinder

therefore,

$$\delta V_3 = \frac{pV}{K}$$ (8.11)

8.8.5

From equations (8.8), (8.10) and (8.11), the additional volume of liquid that is required to be pumped in to raise the internal pressure of the cylinder by p =

$$\delta V = \delta V_1 + \delta V_2 + \delta V_3$$

or,

$$\delta V = \pi R^2 L\varepsilon_L + 2\pi R^2 L\varepsilon_H + pV/K \tag{8.12}$$

8.9.1   EXAMPLE 8.5   ADDITIONAL LIQUID REQUIRED TO RAISE THE PRESSURE INSIDE A CYLINDER

A thin-walled circular cylinder of internal diameter 10 m, of wall thickness 2 cm and of length 5 m is just filled with water. Determine the additional water that is required to be pumped into the vessel to raise its pressure by 0.5 MPa.

$$E = 2 \times 10^{11} \text{ N/m}^2$$
$$v = 0.3$$
$$K = 2 \times 10^9 \text{ N/m}^2$$

8.9.2

From equation (8.8):

$$\delta V_1 = \pi R^2 L\varepsilon_L$$

but,

$$\varepsilon_L = \frac{1}{E}(\sigma_L - v\sigma_H)$$

$$= \frac{pR}{Et}(\tfrac{1}{2} - v)$$

$$= \frac{0.5 \times 10^6 \times 5 \times 0.2}{2 \times 10^{11} \times 2 \times 10^{-2}}$$

or,

$$\varepsilon_L = 1.25 \times 10^{-4}$$

therefore

$$\delta V_1 = \pi \times 25 \times 5 \times 1.25 \times 10^{-4}$$

$$\delta V_1 = 0.049 \text{ m}^3 \tag{8.13}$$

8.9.3

From equation (8.10):

$$\delta V_2 = 2\pi R^2 L\varepsilon_H$$

but,

$$\varepsilon_H = \frac{1}{E}(\sigma_H - v\sigma_L)$$

$$= \frac{pR}{Et}(1 - v/2)$$

$$= \frac{0.5 \times 10^6 \times 5 \times 0.85}{2 \times 10^{11} \times 2 \times 10^{-2}}$$

or.

$$\underline{\varepsilon_H = 5.313 \times 10^{-4}}$$

therefore

$$\delta V_2 = 2\pi \times 25 \times 5 \times 5.313 \times 10^{-4}$$

$$\underline{\delta V_2 = 0.417 \text{ m}^3} \tag{8.14}$$

8.9.4

From equation (8.11):

$$\delta V_3 = \frac{0.5 \times 10^6 \times \pi R^2 L}{K}$$

$$= \frac{0.5 \times 10^6 \times \pi \times 25 \times 5}{2 \times 10^9}$$

$$\underline{\delta V_3 = 0.098 \text{ m}^3} \tag{8.15}$$

8.9.5

From equations (8.13) to (8.15), the additional volume of water required to be pumped into the vessel to raise its internal pressure by 0.5 MPa =

$$\delta V = \delta V_1 + \delta V_2 + \delta V_3$$

$$= 0.049 + 0.417 + 0.098$$

$$\underline{\delta V = 0.564 \text{ m}^3} \tag{8.16}$$

From equation (8.16), it can be seen that the bulk of the additional water required to raise the pressure was because of the radial expansion of the cylinder.

8.10.1   Additional Liquid Required to Raise the Internal Pressure of a Thin-walled Spherical Shell (Based on Small Deflection Elastic Theory)

Assume that the spherical shell is just filled with liquid, and let,

w = the radial deflection of the sphere due to an internal pressure increase of p, as shown in Fig. 8.13.

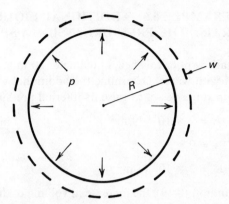

Fig. 8.13.   Deflected form of spherical shell.

The additional liquid that is required to be pumped into the vessel will be because of the following:

(a) Swelling of the structure, due to the application of p.
(b) Compressibility of the liquid itself.

$$\varepsilon = \text{membrane strain due to } p$$

$$\varepsilon = \frac{2\pi(R + w) - 2\pi R}{2\pi R} = \frac{w}{R} \tag{8.17}$$

but,

$$\varepsilon = \frac{1}{E}(\sigma - v\sigma) = \frac{pR}{2tE}(1 - v) \tag{8.18}$$

8.10.2

Change in volume, due to swelling of the shell =

$$\delta V_S = 4\pi R^2 w = 4\pi R^3 \varepsilon$$

$$= 4\pi R^3 \cdot \frac{pR}{2tE}(1 - v)$$

or,

$$\delta V_S = \frac{2\pi p R^4 (1 - v)}{Rt} \tag{8.19}$$

8.10.3

The compressibility of the liquid can be calculated from:

$$\delta V_L = \frac{pV}{K} \tag{8.20}$$

where
$$\delta V_L = \text{change in the volume of the liquid due to compression}$$
$$V = \text{internal volume of sphere}$$
$$= \tfrac{4}{3}\pi R^3$$

8.11.1 EXAMPLE 8.6 ADDITIONAL LIQUID REQUIRED TO RAISE THE PRESSURE INSIDE A SPHERICAL SHELL

A thin-walled spherical shell, of diameter 10 m and a wall thickness 2 cm, is just filled with water. Determine the additional water that is required to be pumped into the vessel to raise its internal pressure by 0.5 MPa.

$$E = 2 \times 10^{11} \text{ N/m}^2$$

$$v = 0.3$$

$$K = 2 \times 10^9 \text{ N/m}^2$$

From equation (8.19), the increase in volume of the spherical shell, due to its swelling under pressure =

$$\delta V_S = \frac{2\pi \times 0.5 \times 10^6 \times 5^4 (0.7)}{2 \times 10^{11} \times 2 \times 10^{-2}}$$

$$\underline{\delta V_S = 0.344 \text{ m}^3} \tag{8.21}$$

From equation (8.20), the additional volume of water to be pumped in, due to the compressibility of the water =

$$\delta V_L = 0.5 \times 10^6 \times \tfrac{4}{3} \times \pi \times 5^3 \times \frac{1}{2 \times 10^9}$$

$$\underline{= 0.131 \text{ m}^3} \tag{8.22}$$

i.e. The total additional quantity of water that is required to be pumped in to raise the pressure =

$$\delta V = \delta V_S + \delta V_L$$
$$\underline{\delta V = 0.475 \text{ m}^3}$$

From equations (8.21) and (8.22), it can be seen that for this spherical shell, the bulk of the additional liquid that was required to raise the pressure by 0.5 MPa was due to the swelling of the shell.

8.12.1 BENDING STRESSES IN CIRCULAR CYLINDERS UNDER UNIFORM PRESSURE

The theory presented in this chapter neglects the effect of bending stresses at the edges of the cylinder, where the vessel may be firmly clamped, as shown in Fig. 8.14.

8.12.2

Fig. 8.15 shows the theoretical and experimental values [12, 13] for the radial deflection of a thin-walled steel cylinder clamped at its ends. The theory was based on the solution of a fourth-order shell differential equation, which is beyond the scope of this book.

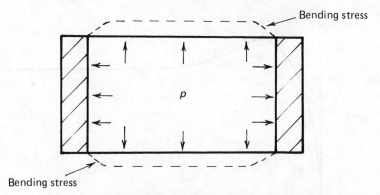

Fig. 8.14.   Deflected form of cylinder under internal pressure.

8.12.3

The vessel was firmly clamped at its ends, but free to move longitudinally, and it had the following properties:

Internal diameter   = 26.04 cm
Wall thickness   = 0.206 cm
Length of cylindrical shell = L = 25.4 cm
External pressure   = 0.6895 MPa
$E = 1.93 \times 10^{11}$ N/m^2
$v = 0.3$ assumed
Initial out-of-roundness   = 0.0102 cm

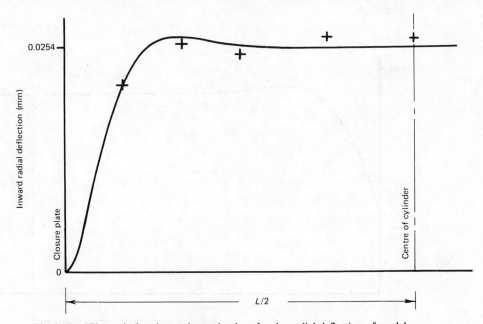

Fig. 8.15.   "Theoretical and experimental values for the radial deflection of model No. 7 at 0.6895 MPa.

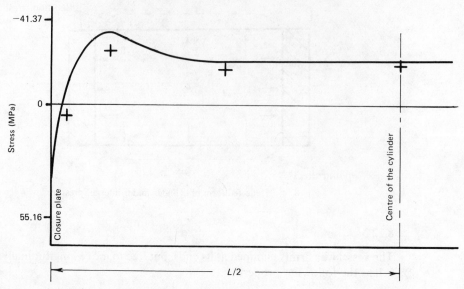

Fig. 8.16.   Longitudinal stress of the outermost fibre at 0.6895 MPa (model No. 7).

8.12.4

Plots of the theoretical and experimental stresses are shown in Figs 8.16 to 8.19, from one end of the vessel (closure plate) to its mid-span.

From Figs 8.15 to 8.19, it can be seen that the effects of bending are very localised.

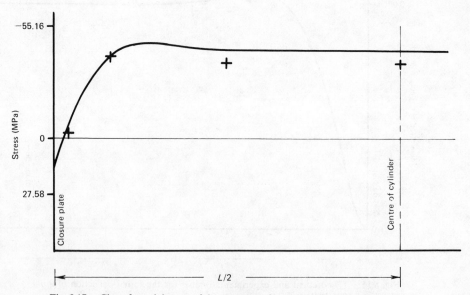

Fig. 8.17.   Cicumferential stress of the outermost fibre at 0.6895 MPa (model No. 7).

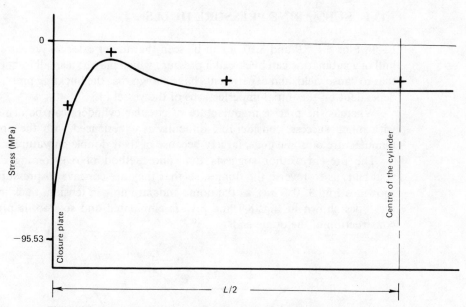

Fig. 8.18.   Longitudinal stress of the innermost fibre at 0.6895 MPa (model No. 7).

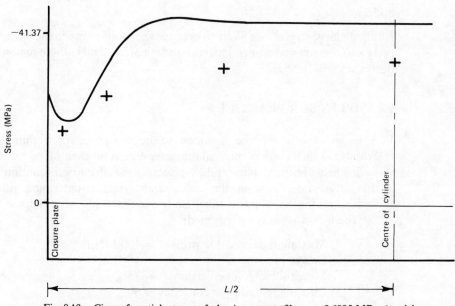

Fig. 8.19.   Circumferential stress of the innermost fibre at 0.6895 MPa (model No. 7).

8.13.1 SUBMARINE PRESSURE HULLS

From Figs 8.1, 8.9 and 8.10, it can be seen that under external pressure, the hull of a submarine can buckle at a pressure which may be a small fraction of that to cause yield, and experiments have shown that the buckling pressure is dependent on the initial imperfections of the vessel [13].

Whereas the precise manufacture of circular cylinders can be achieved with much success, considerable difficulty is experienced with the precise manufacture of dome ends, largely because of their double curvature.

The present author suggests that one method of overcoming dome instability is to reverse the domes, so that they are concave to pressure, as shown in Fig. 8.20. Thus, as the dome ends are now in tension, buckling of the types shown in Figs 8.9 and 8.10 is eliminated and so, too, is precise construction of the dome ends.

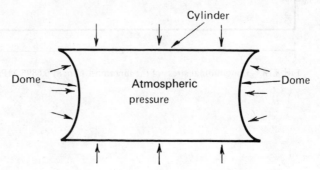

Fig. 8.20. Alternative design of submarine pressure hull.

8.13.2

Although large stresses are likely to occur at the joints between the cylindrical body and the inverted domes, local thickening of the shell in these regions can decrease these stresses.

EXAMPLES FOR PRACTICE 8

1. A boiler, which may be assumed to be composed of a thin-walled cylindrical shell body of internal diameter 4 m, is blocked off by two thin-walled hemispherical dome ends. Neglecting the effects of discontinuity at the intersection between the dome and cylinder, determine suitable thicknesses for the cylindrical shell body and the hemispherical dome ends.

    The following may be assumed;

Maximum permissible stress	$= 100 \text{ MN/m}^2$
Design pressure	$= 1 \text{ MPa}$
Longitudinal joint efficiency	$= 75\%$
Circumferential joint efficiency	$= 50\%$

{Cylinder $t = 2.67$ cm; Dome $t = 2$ cm}

2. If the vessel of Example 1 is just filled with water, determine the additional water that is required to be pumped in, to raise the pressure by 1 MPa.
   The following may be assumed to apply:

   Length of cylindrical portion of vessel = 6 m

   $E = 2 \times 10^{11} \text{ N/m}^2 \quad v = 0.3 \quad K = 2 \times 10^9 \text{ N/m}^2$

   {0.12 m³}

3. A copper pipe of internal diameter 1.25 cm and of wall thickness 0.16 cm is to transport water from a tank that is situated 30 m above it.
   Determine the maximum stress in the pipe, given the following:

   Density of water = 1000 kg/m³

   $g = 9.81 \text{ m/s}^2$

   {11.5 MN/m²}

4. What would be the change in diameter of the pipe of Example 3, due to the applied head of water.

   $\left. \begin{array}{l} E = 1 \times 10^{11} \text{ N/m}^2 \\ v = 0.33 \end{array} \right\}$ for copper

   {1.2 micrometres}

5. A thin-walled spherical pressure vessel of 1 m internal diameter is fed by a pipe of internal diameter 3 cm and of wall thickness 0.16 cm.
   Assuming that the material of construction of the spherical pressure vessel has a yield stress of 0.7 of that of the pipe, determine the wall thickness of the spherical shell.

   {3.8 cm}

6. A spherical pressure vessel of internal diameter 2 m is constructed by bolting together two hemispherical domes with flanges.
   Assuming that the number of bolts used to join the two hemispheres together is twelve, determine the wall thickness of the dome and the diameter of the bolts, given the following:

   (a) Maximum applied pressure   = 0.7 MPa
   (b) Permissible stress in spherical shell =   50 MPa
   (c) Permissible stress in bolts   = 200 MPa

   {t = 0.7 cm; d = 3.4 cm}

7. A thin-walled circular cylinder, blocked off by inextensible end plates, contains a liquid under zero gauge pressure. Show that the additional liquid that is required to be pumped into the vessel, to raise its internal gauge pressure by p, is the same under the following two conditions:

   (a) when axial movement of the cylinder is completely free;
   (b) when the vessel is totally restrained from axial movement.

   It may be assumed that Poisson's ratio (v) for the cylinder material = 0.25.

9

Energy Methods

9.1.1

Energy methods in structural mechanics are some of the most useful methods of theoretical analysis, as they lend themselves to computer solutions of complex structural problems [5, 14], and they can also be extended to computer solutions of many other problems in engineering science [15-22].

There are many energy theorems and principles [14], but only the most popular methods will be considered here, and applications of some of these will be made to practical problems.

9.2.1 THE METHOD OF MINIMUM POTENTIAL (Rayleigh–Ritz)

This states that to satisfy the elasticity and equilibrium equations of an elastic body, the derivative of the total potential, with respect to the displacements, must be zero, i.e.

$$\frac{\partial \pi_p}{\partial u} = 0 \tag{9.1}$$

where,

π_p = total potential
u = displacement

9.2.2 THE PRINCIPLE OF VIRTUAL WORK

This states that if an elastic body under a system of external forces is given a small virtual displacement, then the increase of external virtual work done to the body is equal to the increase in internal virtual strain energy stored in the body.

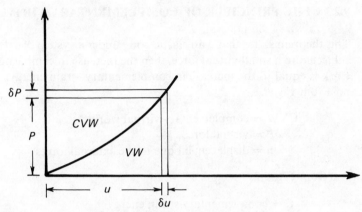

Fig. 9.1.   Force–displacement relationship.

Consider Figs 9.1 and 9.2, where,

δu = a small virtual displacement
$\delta \varepsilon$ = a small virtual strain, as a result of δu
P = external force
σ = stress, due to P

From Fig. 9.1, the increase in virtual work =

$$d(\text{VW}) = \text{area of vertical trapezium} \qquad (9.2)$$

and from Fig. 9.2, the increase in virtual strain energy =

$$d(U_e) = \text{area of vertical trapezium} \qquad (9.3)$$

Equating (9.2) and (9.3):

$$\underline{P * \delta u = \sigma * \delta \varepsilon * \text{volume of body}} \qquad (9.4)$$

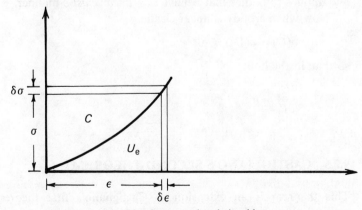

Fig. 9.2.   Stress–strain relationship.

9.2.3 THE PRINCIPLE OF COMPLEMENTARY VIRTUAL WORK

This theorem states that if an elastic body under a system of external forces is subjected to a small virtual force, then the increase in complementary virtual work is equal to the increase in complementary strain energy.

From Fig. 9.1,

CVW = complementary virtual work
δP = virtual force
u = displacement due to the external forces

and from Fig. 9.2,

C = complementary strain energy
$\delta \sigma$ = virtual stress, due to δP
ε = strain due to the external forces

Hence, from Fig. 9.1, the increase in complementary virtual work =

$$d(CVW) = \text{area of horizontal trapezium}$$
$$= \delta P * u \tag{9.5}$$

Similarly, from Fig. 9.2, the increase in complementary strain energy =

$$d(C) = \text{area of horizontal trapezium}$$
$$= \delta \sigma * \varepsilon \tag{9.6}$$

Equating (9.5) and (9.6):

$$\underline{\delta P * u = \delta \sigma * \varepsilon * \text{volume of the body}} \tag{9.7}$$

9.2.4 CASTIGLIANO'S FIRST THEOREM

This is really an extension to the principle of complementary virtual work, and applies to bodies that behave in a *linear elastic* manner.

Now when a body is linear elastic,

$$\delta(C) = \delta(U_e) = \delta P * u$$

so that, in the limit,

$$\frac{\partial U_e}{\partial P} = u \tag{9.8}$$

9.2.5 CASTIGLIANO'S SECOND THEOREM

This theorem is an extension of Castigliano's first theorem, and it is particularly suitable for analysing statically indeterminate frameworks.

Castigliano's second theorem simply states that for a framework with redundant members or forces:

$$\frac{\partial U_e}{\partial R} = \lambda \tag{9.9}$$

where,

U_e = the strain energy of the whole frame
λ = initial lack of fit of the members of the framework
R = the force in any redundant member, to be determined

If there is no initial lack of fit (i.e. the members of the framework have been made precisely), then,

$$\frac{\partial U_e}{\partial R} = 0$$

For most *pin-jointed frames*, the loads are axial; hence the strain energy is given by

$$U = \sum_{i=1}^{N} \frac{P_i^2 l_i}{2 A_i E_i}$$

$$\frac{\partial U}{\partial R} = \frac{\partial U}{\partial P_i} \cdot \frac{\partial P_i}{\partial R}$$

therefore

$$\frac{\partial U}{\partial R} = \sum_{i=1}^{N} \frac{P_i l_i}{A_i E_i} \cdot \frac{\partial P_i}{\partial R} = \text{initial lack of fit}$$

$$= \text{zero (for most undergraduate problems)} \tag{9.10}$$

Thus, by applying equation (9.10) to each redundant member of "force", in turn, the required number of simultaneous equations is obtained, and hence, the unknown redundant "forces" can be determined. Once the redundant "forces" are known, the other "forces" can be determined through statics.

9.3.1   STRAIN ENERGY STORED IN A ROD UNDER AXIAL LOADING

Consider the load–displacement relationship of a uniform section rod, as shown in Fig. 9.3.

Now when the rod displaces by u, the force required to achieve this displacement is P. Furthermore, as the average force during this load–displacement relationship is $P/2$, the

Work done = the shaded area of Fig. 9.3

$$= \frac{P}{2} * u \tag{9.11}$$

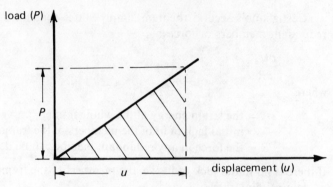

Fig. 9.3. Load–displacement relationship for a rod.

However, this work done will be stored by the rod in strain energy U_e, so that,

$$U_e = \tfrac{1}{2}P * u \tag{9.12}$$

but,

$$P = \sigma * A \tag{9.13}$$

and,

$$u = \varepsilon * l = \sigma * l/E \tag{9.14}$$

where,

σ = stress due to P
ε = strain due to P
A = cross-sectional area of rod
l = length of rod

Substituting equations (9.13) and (9.14) into (9.12):

$$U_e = \frac{\sigma^2}{2E} * \text{volume of the rod} \tag{9.15}$$

For an element of the rod, of length "dx",

$$d(U_e) = \frac{\sigma^2}{2E} * A * dx \tag{9.16}$$

where $d(U_e)$ = strain energy in the rod élement.

9.3.2 STRAIN ENERGY STORED IN A BEAM, SUBJECTED TO COUPLES OF MAGNITUDE M AT ITS ENDS

Consider the element of the beam of length "dx" and assume it is subjected to two end couples, each of magnitude M, as shown in Fig. 9.4.

The work done in bending the element "dx" of the beam of Fig. 9.4 =

$$\text{W.D.} = \tfrac{1}{2}M\theta$$

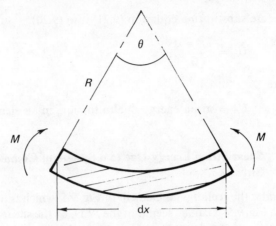

Fig. 9.4.   Element of beam in pure bending.

but,

$$\theta = dx/R$$

therefore

$$\text{W.D.} = \tfrac{1}{2}M * dx/R \tag{9.17}$$

Now from equation (4.1):

$$\frac{1}{R} = \frac{M}{EI} \tag{9.18}$$

Substituting equation (9.18) into (9.17),

$$\text{W.D.} = d(U_b) = \frac{M^2}{2EI} * dx \tag{9.19}$$

where,

$$d(U_b) = \text{bending strain energy in the beam element}$$

9.3.3   STRAIN ENERGY STORED IN A UNIFORM CIRCULAR SECTION SHAFT, OWING TO A TORQUE T

Let,

$$\theta = \text{angle of twist of one end of the shaft relative to the other}$$
$$T = \text{applied torque}$$

therefore

$$\text{W.D.} = U_T = \tfrac{1}{2}T * \theta \tag{9.20}$$

However, from equation (6.5):

$$\theta = T * dx/(G * J) \tag{9.21}$$

therefore substituting equation (9.21) into (9.20):

$$U_T = \frac{T^2 * dx}{GJ} \qquad (9.22)$$

where,

U_T = strain energy due to torsion, in an element "dx"

9.3.4   Shear Strain Energy Due to a System of Complementary Shear Stresses τ

Consider the rectangular element of Fig. 9.5, which is in a state of pure shear. From Fig. 9.5, it can be seen that the W.D. by the shear stresses τ, in changing the shape of the element =

$$\text{W.D.} = U_s = \frac{\tau}{2} * (t * dx) * (dy * \gamma)$$

but,

$$\gamma = \tau/G$$

therefore

$$U_s = \frac{\tau^2}{2G} * \text{volume} \qquad (9.23)$$

where,

U_s = shear strain energy

volume = $t * dx * dy$

t = *thickness of plate*

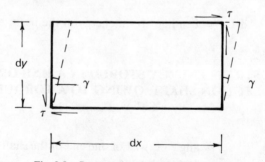

Fig. 9.5.   Rectangular element in pure shear.

9.4.1   EXAMPLE 9.1   RODS UNDER AXIAL LOADING

A rod consists of three elements, each of length 0.5 m, joined firmly together. If this rod is subjected to an axial tensile force of 1 MN, determine the total extension of the rod, using strain energy principles, and given that the cross-sectional areas of the elements are

Section ①: $A_1 = 5E\text{-}3 \ m^2$

Section ②: $A_2 = 3E\text{-}3 \ m^2$

Section ③: $A_3 = 1E\text{-}2 \ m^2$

Elastic modulus = $1 \times 10^{11} \ N/m^2$

9.4.2

Section ①: $\sigma_1 = \dfrac{1 \ MN}{5E\text{-}3 \ m^2} = \underline{200 \ MN/m2}$

Section ②: $\sigma_2 = \dfrac{1 \ MN}{3E\text{-}3 \ m^2} = \underline{333.3 \ MN/m^2}$

Section ③: $\sigma_3 = \dfrac{1 \ MN}{1E\text{-}2 \ m^2} = \underline{100 \ MN/m^2}$

Let δ = total deflection of the rod, so that,

$$\text{W.D.} = \tfrac{1}{2} * 1 \ MN * \delta \tag{9.24}$$

From equation (9.15):

$$U_e = \frac{\sigma_1^2}{2E} * A_1 * l + \frac{\sigma_2^2}{2E} * A_2 * l + \frac{\sigma_3^2}{2E} * A_3 * l \tag{9.25}$$

Equating (9.24) and (9.25):

$$0.5\delta = \frac{0.5}{2E5} (200^2 \times 5E\text{-}3 + 333.3^2 \times 3E\text{-}3 + 100^2 \times 1E\text{-}2)$$

or,

$$\delta = \frac{1}{2E5} (200 + 333.3 + 100)$$

$$\underline{\delta = 3.17 \ mm}$$

9.5.1   EXAMPLE 9.2   DEFLECTION OF A CANTILEVER

Determine the maximum deflection of the end loaded cantilever of Fig. 9.6, using strain energy principles.

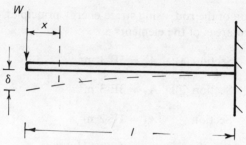

Fig. 9.6.   End loaded cantilever.

9.5.2

Let,

I = second moment of area of the beam section, about a horizontal axis.

W.D. by the load = $\frac{1}{2}W * \delta$ (9.26)

U_b = bending strain energy in the beam

$$= \int_0^l \frac{M^2}{2EI}\,dx$$

but,

$$M = -Wx$$

therefore

$$U_b = \frac{W}{2EI}\int_0^l x^2\,dx$$

$$= \frac{Wl^3}{6EI}$$ (9.27)

Equating (9.26) and (9.27),

$$\underline{\delta = \frac{Wl^3}{3EI} \text{ (as required)}}$$

9.6.1   EXAMPLE 9.3   DEFLECTION OF A SIMPLY-SUPPORTED BEAM

Determine the deflection at mid-span, for the centrally loaded beam of Fig. 9.7, which is simply-supported at its ends.

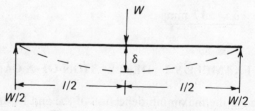

Fig. 9.7.   Centrally loaded beam.

9.6.2

$$\text{W.D. by load} = \tfrac{1}{2}W\delta \tag{9.28}$$

$$U_b = \int \frac{M^2}{2EI}\, dx$$

$$= \frac{1}{2EI} \int_0^{l/2} \left(\frac{Wx}{2}\right)^2 dx * 2 \tag{9.29}$$

$$= \frac{W^2}{4EI} \left[\frac{x^3}{3}\right]_0^{l/2}$$

or,

$$U_b = \frac{W^2 l^3}{96EI} \tag{9.30}$$

Equating (9.28) and (9.30):

$$\delta = \frac{Wl^3}{48EI} \text{ (as required)}$$

N.B. It should be noted that in equation (9.29), the upper limit of the integral was $l/2$. It was necessary to integrate in this manner, because the expression for the bending moment, namely $M = Wx/2$, applied only between $x = 0$ and $x = l/2$.

9.7.1   DEFLECTION OF THIN CURVED BEAMS

In this theory, it will be assumed that the effects of shear strain energy and axial strain energy are negligible, and that all the work done by external loads on these beams is in fact absorbed by them in the form of bending strain energy.

Castigliano's theorems will be used in this section, and the method will be demonstrated by applying it to a number of examples.

Castigliano's first theorem requires a *load to act in the direction of the required displacement*; when the value of such a displacement is to be determined, at a point where no load points in the same direction, it will be necessary to assume an imaginary load to act in the direction of the required displacement, and, later, to set this imaginary load to zero.

9.8.1   EXAMPLE 9.4   DEFLECTION OF A QUADRANT

A thin curved beam is in the form of a quadrant of a circle, as shown in Fig. 9.8. One end of this quadrant is firmly fixed to a solid base, whilst the other end is subjected to a point load *W*, acting vertically downwards. Determine expressions for the vertical and horizontal displacements at the free end.

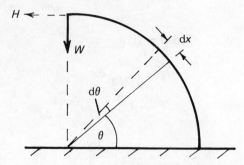

Fig. 9.8. Thin curved beam (quadrant).

9.8.2

As the value of the horizontal displacement is required to be determined at the free end, and as there is no load acting in this direction, it will be necessary to assume *an imaginary load H*, as shown by the dashed line. Later in the calculation, it will be necessary to note that $H = 0$.

Considered an element "dx" at any given angle θ from the base. At this point the bending moment =

$$M = -WR \cos \theta - HR(1 - \sin \theta)$$

(N.B. The *sign convention* for bending moment is not important when using Castigliano's first theorem.)

Let,

δV = downward vertical displacement at the free end
δH = horizontal deflection (to the left) at the free end

From equation (9.19), the bending strain energy of the element "dx" =

$$d(U_b) = \frac{M^2}{2EI} dx$$

or,

$$U_b = \int \frac{M^2}{2EI} dx$$

From Castigliano's first theorem,

$$\delta V = \frac{\partial U_b}{\partial W} = \frac{\partial U_b}{\partial M} \frac{\partial M}{\partial W} = \frac{1}{EI} \int M \frac{\partial M}{\partial W} dx \qquad (9.31)$$

therefore

$$\delta V = \frac{1}{EI} \int_0^{\pi/2} [WR \cos \theta + HR(1 - \sin \theta)]R \cos \theta R \, d\theta$$

but,

$$\underline{H = 0}$$

therefore

$$\delta V = \frac{WR^3}{EI} \int_0^{\pi/2} \cos^2 \theta \, d\theta = \frac{WR^3}{EI} \int_0^{\pi/2} \frac{(1 + \cos 2\theta)}{2} \, d\theta$$

$$\underline{\delta V = \pi WR^3/(4EI)} \qquad\qquad\qquad (9.32)$$

Similarly,

$$\delta H = \frac{\partial U_b}{\partial H} = \frac{\partial U_b}{\partial M} \frac{\partial M}{\partial H} = \frac{1}{EI} \int M \frac{\partial M}{\partial H} \, dx$$

$$= \frac{1}{EI} \int_0^{\pi/2} [WR \cos \theta + HR(1 - \sin \theta)]R(1 - \sin \theta)R \cdot d\theta$$

but,

$$\underline{H = 0}$$

therefore

$$\delta H = \frac{WR^3}{EI} \int_0^{\pi/2} (\cos \theta - \sin \theta \cos \theta) \, d\theta$$

$$= \frac{WR^3}{EI} \left[\sin \theta - \frac{\sin^2 \theta}{2} \right]_0^{\pi/2}$$

$$\underline{\delta H = WR^3/(2EI)} \qquad\qquad\qquad (9.33)$$

Thus, it can be seen that, owing to W, there is a horizontal displacement in addition to a vertical displacement. The value of the resultant displacement can be obtained by considerations of Pythagoras' theorem, together with elementary trigonometry.

9.9.1   EXAMPLE 9.5   DEFLECTION OF A SEMI-CIRCULAR CURVED BEAM

A thin curved beam, which is in the form of a semi-circle, is firmly fixed at its base, and it is subjected to point loads at its free end, as shown in Fig. 9.9. Determine expressions for the vertical and horizontal displacements at its free end.

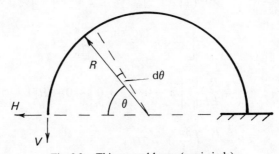

Fig. 9.9.   Thin curved beam (semi-circle).

9.9.2

Let,

$$\delta V = \text{deflection in the direction of } V$$
$$\delta H = \text{deflection in the direction of } H$$

At any angle θ, the bending moment acting on an element of the beam, of length $R \, d\theta =$

$$M = VR(1 - \cos \theta) - HR \sin \theta$$

Now,

$$\delta V = \frac{1}{EI} \int M \, \frac{\partial M}{\partial V} \cdot R \, d\theta$$

and,

$$\delta H = \frac{1}{EI} \int M \, \frac{M}{H} \cdot R d\theta$$

therefore

$$\delta V = \frac{1}{EI} \int [VR(1 - \cos \theta) - HR \sin \theta] R(1 - \cos \theta) R \, d\theta$$

$$= \frac{R^3}{EI} \int \{V(1 - 2 \cos \theta + \cos^2 \theta) - H(\sin \theta - \sin \theta \cos \theta)\} \, d\theta$$

$$= \frac{R^3}{EI} \int V \left[1 - 2 \cos \theta + \frac{(1 + \cos 2\theta)}{2} \right]$$

$$- H(\sin \theta - \sin \theta \cos \theta) \, d\theta$$

$$= \frac{R^3}{EI} \left\{ V \left[\theta - 2 \sin \theta + \frac{\theta}{2} + \frac{\sin 2\theta}{4} \right] \right.$$

$$\left. - H \left[- \cos \theta - \frac{\sin^2 \theta}{2} \right] \right\}_0^\pi$$

$$= \frac{R^3}{EI} \left\{ V \left[\pi - 0 + \frac{\pi}{2} + 0 \right] - 0 - H[1 - 0] + H[-1 - 0] \right\}$$

$$\delta V = \frac{R^3}{EI} \left(\frac{3\pi}{2} V - 2H \right) \tag{9.34}$$

$$\delta H = \frac{1}{EI} \int M \frac{\partial M}{\partial H} R \, d\theta$$

$$= \frac{1}{EI} \int [VR(1 - \cos \theta) - HR \sin \theta](-R \sin \theta)R \, d\theta$$

$$= \frac{R^3}{EI} \left\{ \int [-\sin \theta + \sin \theta \cos \theta]V + H \sin^2 \theta \right\} d\theta$$

$$= \frac{R^3}{EI} \left\{ \left[\cos \theta + \frac{\sin^2 \theta}{2} \right] V + \int H \frac{(1 - \cos 2\theta)}{2} \, d\theta \right\}$$

$$= \frac{R^3}{EI} \left\{ \left[\cos \theta + \frac{\sin^2 \theta}{2} \right] V + \frac{H}{2} \left[\theta - \frac{\sin 2\theta}{2} \right] \right\}_0^\pi$$

$$= \frac{R^3}{EI} \left[(-1 + 0) - (1 + 0)]V + \frac{H}{2} [(\pi - 0) - (0 - 0)] \right]$$

$$= \frac{R^3}{EI} \left(-2V + \frac{\pi H}{2} \right) \tag{9.35}$$

9.10.1   EXAMPLE 9.6   DEFLECTION OF A BEAM THAT IS PART CURVED AND PART STRAIGHT

Determine expressions for the deflections at the free end of the thin curved beam shown in Fig. 9.10.

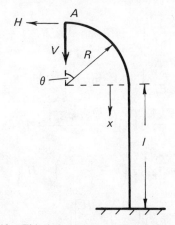

Fig. 9.10.   Thin beam (part curved and part straight).

9.10.2

For convenience, this beam will be considered in two sections, i.e. a curved section and a straight section.

Let,

$$\delta \overset{\bullet}{V} = \text{deflection at the free end in the direction of } V$$
$$= \delta V_{\text{c}} + \delta V_{\text{s}}$$

$$\delta H = \text{deflection at free end in the direction of } H$$
$$= \delta H_{\text{c}} + \delta H_{\text{s}}$$

where,

δV_{c} = component of deflection at the free end in the direction of V, due to the curved section

δH_{c} = component of deflection at the free end in the direction of H, due to the curved section

δV_{s} = component of deflection at the free end in the direction of V, due to the straight section

δH_{s} = component of deflection at the free end in the direction of H, due to the straight section

As the problem is a linear elastic one, it is perfectly acceptable to add together the appropriate components of δV and δH.

9.10.3   Curved Section

At any angle θ, the bending moment =

$$M = VR \sin \theta + HR(1 - \cos \theta)_1$$

and

$$\delta V_{\text{c}} = \frac{1}{EI} \int_0^{\pi/2} M \cdot \frac{\partial M}{\partial V} R \, d\theta$$

$$= \frac{1}{EI} \int_0^{\pi/2} [VR \sin \theta + HR(1 - \cos \theta)] \cdot R \sin \theta \cdot R \, d\theta$$

$$= \frac{R^3}{EI} \int_0^{\pi/2} [V \sin^2 \theta + H(\sin \theta - \sin \theta \cos \theta)] \, d\theta$$

$$= \frac{R^3}{EI} \int \left[V \frac{(1 - \cos 2\theta)}{2} + H(\sin \theta - \sin \theta \cos \theta) \right] d\theta$$

$$= \frac{R^3}{EI} \left[\frac{V}{2} \left(\theta - \frac{\sin 2\theta}{2} \right) + H \left(- \cos \theta - \frac{\sin^2 \theta}{2} \right) \right]_0^{\pi/2}$$

$$\delta V_{\text{c}} = \frac{R^3}{EI} \left[\left(\frac{\pi V}{4} \right) + \frac{H}{2} \right] \tag{9.36}$$

Similarly,

$$\delta H_c = \frac{1}{EI} \int_0^{\pi/2} M \cdot \frac{\partial M}{\partial H} \cdot R \, d\theta$$

$$= \frac{1}{EI} \int [VR \sin \theta + HR(1 - \cos \theta)] \cdot R(1 - \cos \theta) \cdot R \, d\theta$$

$$= \frac{R^3}{EI} \int [V \sin \theta (1 - \cos \theta) + H(1 - \cos \theta)^2] \, d\theta$$

$$= \frac{R^3}{EI} \int [V(\sin \theta - \sin \theta \cos \theta) + H(1 - 2 \cos \theta + \cos^2 \theta)] \, d\theta$$

$$= \frac{R^3}{EI} \left\{ V\left(-\cos \theta - \frac{\sin^2 \theta}{2} \right) \right.$$

$$\left. + \int H\left[1 - 2 \cos \theta + \frac{(1 - \cos 2\theta)}{2} \right] d\theta \right\}_0^{\pi/2}$$

$$= \frac{R^3}{EI} \left[V\left(-\cos \theta - \frac{\sin^2 \theta}{2} \right) \right.$$

$$\left. + H\left(\theta - 2 \sin \theta + \frac{\theta}{2} - \frac{\sin 2\theta}{4} \right) \right]_0^{\pi/2}$$

$$\delta H_c = \frac{R^3}{EI} \left[\frac{V}{2} + H\left(\frac{3\pi}{4} - 2 \right) \right] \qquad (9.37)$$

9.10.4 Straight Section

At any distance x, the bending moment =

$$M = VR + H(R + X)$$

Now,

$$\delta V_s = \frac{1}{EI} \int_0^l M \frac{\partial M}{\partial V} \cdot dx$$

$$= \frac{1}{EI} \int_0^l [VR + H(R + x)]R \, dx = \frac{R}{EI} \left[VRx + HRx + \frac{Hx^2}{2} \right]$$

$$\delta V_s = \frac{R}{EI} \left[VRl + H\left(Rl + \frac{l^2}{2} \right) \right] \qquad (9.38)$$

Similarly,

$$\delta H_s = \frac{1}{EI} \int_0^l [VR + H(R + x)](R + x)\,dx$$

$$= \frac{1}{EI} \left\{ VR^2 x + \frac{VRx^2}{2} + H\left(R^2 x + Rx^2 + \frac{x^3}{3} \right) \right\}_0^l$$

therefore

$$\delta H_s = \frac{1}{EI} \left\{ VRl\left(R + \frac{l}{2} \right) + Hl\left(R^2 + Rl + \frac{l^2}{3} \right) \right\} \tag{9.39}$$

From equations (9.36) to (9.39):

$$\delta V = \frac{1}{EI} \left\{ R^3 \left[\frac{\pi V}{4} + \frac{H}{2} \right] + R[VRl + Hl(R + l/2)] \right\} \tag{9.40}$$

$$\delta H = \frac{1}{EI} \left\{ \frac{R^3}{EI} \left[\frac{V}{2} + H\left(\frac{3\pi}{4} - 2 \right) \right] + VRl\left(R + \frac{l}{2} \right) \right.$$

$$\left. + Hl\left(R^2 + Rl + \frac{l^2}{3} \right) \right\} \tag{9.41}$$

9.11.1   EXAMPLE 9.7   OUT-OF-PLANE DEFLECTION OF A THIN CURVED BEAM

Determine the deflection at the free end of the thin curved beam shown in Fig. 9.11, which has an out-of-plane concentrated load applied to its free end.

9.11.2

In this case, the beam is subjected to both bending and torsion; hence, it will be necessary to consider bending strain energy in addition to torsional strain energy.

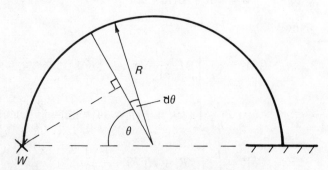

Fig. 9.11.   Thin curved beam under an out-of-plane load.

At any angle θ, the element ($R\,d\theta$) of the beam is subjected to a bending moment M, and a torque T, which are evaluated as follows:

$$M = WR \sin \theta$$

$$T = WR(1 - \cos \theta)$$

From equations (9.19) and (9.22), the total strain energy, U, is given by

$$U = \int \frac{M^2}{2EI}\,dx + \int \frac{T^2}{2GJ}\,dx$$

Let,

δW = out-of-plane deflection under the load W, acting at the free end.

Now,

$$\delta W = \frac{1}{EI} \int M \frac{\partial M}{\partial W} R\,d\theta + \frac{1}{GJ} \int T \frac{\partial T}{\partial W} R\,d\theta$$

$$= \frac{R}{EI} \int WR \sin \theta \cdot R \sin d\theta$$

$$+ \frac{R}{GJ} \int [WR(1 - \cos \theta)]R(1 - \cos \theta)\,d\theta$$

$$= \frac{WR^3}{EI} \int_0^\pi \frac{(1 - \cos 2\theta)}{2}\,d\theta$$

$$+ \frac{WR^3}{GJ} \int_0^\pi [1 - 2 \cos \theta + \cos^2 \theta]\,d\theta$$

$$= WR^3 \int \frac{1}{2}\left[\theta - \frac{\sin 2\theta}{2}\right] \frac{1}{EI}$$

$$+ \frac{1}{GJ} \left[\theta - 2 \sin \theta + \frac{1}{2}\left(\theta + \frac{\sin 2\theta}{2}\right)\right]\Big\}_0^\pi$$

$$= WR^3 \left\{\frac{1}{2EI} [\pi] + \frac{1}{GJ}\left[\left(\pi - 0 + \frac{\pi}{2} + 0\right) - (0)\right]\right\}$$

$$\delta W = WR^3\left(\frac{\pi}{2EI} + \frac{3\pi}{2GJ}\right) \tag{9.42}$$

9.12.1   EXAMPLE 9.8   BENDING MOMENT DISTRIBUTION IN A THIN CURVED RING

Determine expressions for the values of the bending moments at the points "A" and "B" of the thin ring shown in Fig. 9.12, and also obtain expressions for the changes in diameter of the ring in the directions of the applied loads.

Hence, or otherwise, sketch the bending moment diagram, around the circumference of the ring, when $H = 0$.

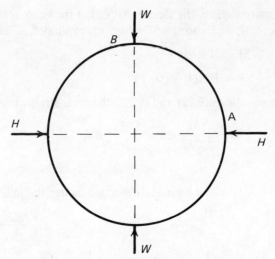

Fig. 9.12. Thin ring, under diametral loads.

9.12.2

Because of symmetry, it is only necessary to consider a quadrant of the ring, as shown in Fig. 9.13.

Let,

M_0 = unknown bending moment at A, which can be assumed to be the redundancy.

M_1 = unknown bending moment at B, which can be obtained from M_0, together with considerations of elementary statics.

As this problem is statically indeterminate, it will be necessary to use Castigliano's second theorem to determine the redundancy M_0.

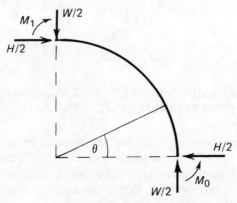

Fig. 9.13. Quadrant of ring.

At angle θ, the bending moment acting on an element, $R \cdot d\theta =$

$$M = M_0 + \frac{WR}{2}(1 - \cos\theta) - \frac{HR}{2}\sin\theta \qquad (9.43)$$

To find M_0

From Castigliano's second theorem:

$$\frac{\partial U_b}{\partial M_0} = \text{initial lack-of-fit} = 0 \text{ (in this case, as the ring is assumed to}$$

be geometrically perfect)

or,

$$\frac{\partial U_b}{\partial M_0} = \frac{\partial U_b}{\partial M} \cdot \frac{\partial M}{\partial M_0}$$

i.e.

$$0 = \frac{1}{EI}\int M \cdot \frac{M}{M_0} \cdot R d\theta$$

$$0 = \frac{4}{EI}\int_0^{\pi/2}\left[M_0 + \frac{WR}{2}(1 - \cos\theta) - \frac{HR}{2}\sin\theta\right] \cdot R \cdot d\theta$$

$$0 = \frac{4}{EI}\int_0^{\pi/2}\left(M_0 + \frac{WR}{2} - \frac{WR\cos\theta}{2} - \frac{HR\sin\theta}{2}\right)R \cdot d\theta$$

$$0 = \left[M_0\theta + \frac{WR}{2}\theta - \frac{WR}{2}\sin\theta + \frac{HR}{2}\cos\theta\right]_0^{\pi/2}$$

$$0 = M_0\frac{\pi}{2} + WR\frac{\pi}{4} - \frac{WR}{2} + 0 + 0 - \frac{HR}{2}$$

$$M_0 = (W + H)\frac{R}{\pi} - \frac{WR}{2} \qquad (9.44)$$

An expression for M_1 can be determined from equation (9.43), by setting $\theta = 90°$, and by the use of equation (9.44), i.e.

$$M_1 = M_0 + \frac{WR}{2}\left(1 - \cos\frac{\pi}{2}\right) - \frac{HR}{2}\sin\frac{\pi}{2}$$

or,

$$M_1 = \frac{WR}{\pi} + HR\left(\frac{1}{\pi} - \tfrac{1}{2}\right) \qquad (9.45)$$

To find the *change in the diameter* of the ring in the direction of W

$$\delta V = \frac{1}{EI} \int M \cdot \frac{\partial M}{\partial W} \cdot dx$$

$$= \frac{4}{EI} \int_0^{\pi/2} \left[M_0 + \frac{WR}{2} (1 - \cos\theta) \right.$$

$$\left. - \frac{HR}{2} \sin\theta \right] \cdot \frac{R}{2} (1 - \cos\theta) \cdot R \cdot d\theta$$

$$= \frac{4R^3}{EI} \int_0^{\pi/2} \left[\frac{W}{\pi} + \frac{H}{\pi} - \frac{W}{2} + \frac{W}{2} - \frac{W}{2} \cos\theta \right.$$

$$\left. - \frac{H}{2} \sin\theta \right] \tfrac{1}{2}(1 - \cos\theta) \cdot d\theta$$

$$= \frac{2R^3}{EI} \left\{ \frac{(W+H)}{\pi} (\theta - \sin\theta) - \frac{W}{2} \sin\theta + \frac{H}{2} \cos\theta \right.$$

$$\left. + \int \left[\frac{W}{2} \cos^2\theta + \frac{H}{2} \sin\theta \cos\theta \right] d\theta \right\}_0^{\pi/2}$$

$$= \frac{2R}{EI} \left\{ \frac{(W+H)}{\pi} \left(\frac{\pi}{2} - 1 \right) - \frac{W}{2} - \frac{H}{2} \right.$$

$$\left. + \int_0^{\pi/2} \left[\frac{W}{2} \frac{(1 + \cos 2\theta)}{2} + \frac{H}{2} \sin\theta \cos\theta \right] \right\} d\theta$$

$$= \frac{2R^3}{EI} \left\{ \frac{(W+H)}{\pi} \left(\frac{\pi}{2} - 1 \right) - \frac{W}{2} - \frac{H}{2} + \frac{W}{4} \left[\theta + \frac{\sin 2\theta}{2} \right]_0^{\pi/2} \right.$$

$$\left. + \frac{H}{4} [\sin^2\theta]_0^{\pi/2} \right\}$$

$$= \frac{2R^3}{EI} \left\{ -\frac{(W+H)}{\pi} + \frac{W\pi}{8} + 0 + \frac{H}{4} \right\}$$

$$\delta V = \frac{2R^3}{EI} \left[W \left(\frac{\pi}{8} - \frac{1}{\pi} \right) + H \left(\tfrac{1}{4} - \frac{1}{\pi} \right) \right] \tag{9.46}$$

By a similar process, the change in diameter of the ring in the direction of H can be found.

$$\delta H = \frac{R^3}{EI} \left[H \left(\frac{\pi}{8} - \frac{1}{\pi} \right) + W \left(\tfrac{1}{4} - \frac{1}{\pi} \right) \right] \tag{9.47}$$

N.B. Changes in the diameters $= 2 * \delta V$ and $2 * \delta H$, respectively.
To determine the bending moment diagram, when $H = 0$

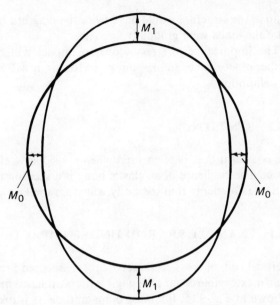

Fig. 9.14.   Bending moment distribution, when $H = 0$.

From equation (9.43),

$$M = \left(\frac{WR}{\pi} - \frac{WR}{2}\right) + \frac{WR}{2}\,(1 - \cos\,\theta)$$

$$M = \frac{WR}{\pi} - \frac{WR}{2}\,\cos\,\theta$$

@ $\theta = \dfrac{\pi}{2}$

$$\underline{M_1 = \frac{WR}{\pi}}$$

@ $\theta = 0$

$$\underline{M_2 = \left(\frac{WR}{\pi} - \frac{WR}{2}\right)}$$

The bending moment diagram is shown in Fig. 9.14.

9.13.1   SUDDENLY APPLIED AND IMPACT LOADS

When structures are subjected to suddenly applied or impact loads, the resulting stresses can be considerably greater than those which will occur if these same loads are gradually applied.

Using a theoretical solution, it will be proven later, that if a stationary load were suddenly applied to a structure, at zero velocity, then the stresses set up in the structure would have twice the magnitude that they would have had if the loads were gradually applied.

The importance of stresses due to impact will be demonstrated by a number of worked examples, but prior to this, it will be necessary to make a few definitions.

9.13.2   RESILIENCE

This is a term that is often used when considering elastic structures under impact. The resilience of an elastic body is a measurement of the amount of elastic strain energy that the body will store under a given load.

9.14.1   EXAMPLE 9.9   ROD UNDER IMPACT

A vertical rod, of cross-sectional area A, is secured firmly at its top end, and has an inextensible collar of negligible mass attached firmly to its bottom end, as shown in Fig. 9.15. If a mass of magnitude M is dropped onto the collar, from a distance h above it, determine the maximum values of stress and deflection that will occur owing to this impact. Neglect energy and other losses, and assume that the mass is in the form of an annular ring.

9.14.2

Energy and friction losses are neglected, so that the rod is assumed to absorb all the energy on impact. From the point of view of the structural analyst, this

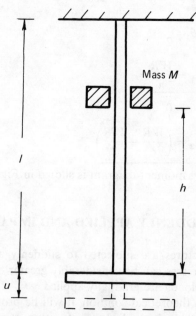

Fig. 9.15.   Rod under impact.

assumption is reasonable, as it simplifies the solution, and the errors are on the so-called "safe side".

Let,

u = maximum deflection of the rod on impact

therefore

P.E. of mass = $Mg(h + u)$
= strain energy stored in the rod

i.e.

$$Mg(h + u) = \frac{\sigma^2}{2E} * \text{volume of rod} \tag{9.48}$$

where,

σ = the maximum stress in the rod, due to impact
P.E. = potential energy

but,

$$u = \varepsilon l = \sigma l / E$$

which on substitution into equation (9.48) gives

$$Mg(h + \sigma l / E) = \frac{\sigma^2}{2E} * Al$$

or,

$$\sigma^2 - \frac{2Mg\sigma}{A} - \frac{2MghE}{Al} = 0$$

therefore

$$\sigma = \frac{Mg}{A} + \sqrt{\frac{M^2 g^2}{A^2} + \frac{2MghE}{Al}} \tag{9.49}$$

and,

$$u = \sigma l / E \tag{9.50}$$

9.14.3   Suddenly Applied Load

If the mass M in Fig. 9.15 were *just above* the top surface of the collar, and *suddenly released*, so that $h = 0$ in equation (9.49), then

$$\sigma = \frac{2Mg}{A}$$

i.e. *a suddenly applied load*, at zero velocity, *induces twice the stress* of a gradually applied load of the same magnitude.

9.15.1

If in Example 9.9, the following applied, determine the maximum stress on impact, and, also, the maximum stress if the same load were gradually applied to the top of the collar:

$$M = 1 \text{ kg} \qquad A = 400 \text{ mm}^2 \qquad l = 1 \text{ m} \qquad h = 0.5 \text{ m}$$
$$E = 2 \times 10^{11} \text{ N/m}^2 \qquad \qquad g = 9.81 \text{ m/s}^2$$

9.15.2

From equation (9.49), Maximum stress on impact =

$$\sigma = \frac{1 \times 9.81}{400 \text{ E-6}} + \sqrt{\frac{1^2 \times 9.81^2}{(400 \text{ E-6})^2} + \frac{2 \times 1 \times 9.81 \times 0.5 \times 2\text{E}11}{400 \text{ E-6} \times 1}}$$

$$\sigma = 0.025 + 70.04 = 70.07 \text{ MN/m}^2$$

If the load were gradually applied to the top of the collar,

$$\text{Maximum static stress} = 0.025 \text{ mN/m}^2$$

i.e. *the impact stress is about* 2800 *times greater than the static stress.*

9.16.1   EXAMPLE 9.10   BEAM UNDER IMPACT

A simply-supported beam of uniform section is subjected to an impact load at mid-span, as shown in Fig. 9.16. Determine expressions for the maximum bending moment and deflection of the beam.

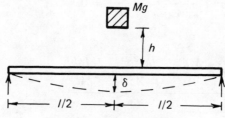

Fig. 9.16.   Beam under impact.

9.16.2

Let,

$$\delta = \text{the maximum central deflection of the beam due to impact}$$
$$W_e = \text{equivalent static load to cause the deflection } \delta$$
$$\text{P.E. of load} = Mg(h + \delta) \tag{9.15}$$

From equation (9.30),

$$U_b = \frac{W_e^2 l^3}{96EI} \tag{9.52}$$

Equating (9.51) and (9.52):

$$Mg(h + \delta) = \frac{W_e^2 l^3}{96EI} \qquad (9.53)$$

but from Section 9.5.1,

$$\delta = \frac{W_e l^3}{48EI},$$

which on substitution into equation (9.53) results in

$$-Mgh - \frac{Mgl^3 W_e}{48EI} + \frac{l^3}{96EI} W_e^2 = 0$$

or,

$$W_e^2 - 2MgW_e - \frac{96EIMgh}{l^3} = 0$$

therefore

$$W_e = Mg + \sqrt{M^2 g^2 + 96EIMgh/l^3} \qquad (9.54)$$

If Mg were suddenly applied, at zero velocity, so that $h = 0$ in equation (9.54), then,

$$W_e = 2Mg$$

i.e. once again, it can be seen that a suddenly applied load, at zero velocity, has twice the value of a gradually applied load.

N.B. $\delta = \dfrac{W_e l^3}{48EI}$

9.17.1   EXAMPLE 9.11   IMPACT ON A BEAM, SUPPORTED BY A WIRE AT ONE END

A uniform section beam, which is initially horizontal, is simply-supported at one end, and is supported at the other end by an elastic wire, as shown in Fig. 9.17. Assuming that a mass of magnitude 2 kg, which is situated at a height of 0.2 m above the beam, is dropped onto the mid-span of the beam, determine the central deflection of the beam and the maximum stress induced in the wire. The beam and wire may be assumed to have negligible masses, and the following may also be assumed to apply:

$E = 2E11$ N/m^2 (for both beam and wire)
$I =$ second moment of area of the beam section, about a
  horizontal plane $= 2E\text{-}8$ m^4
$A =$ cross-sectional area of wire $= 1.3E\text{-}6$ m^2
$L =$ length of beam $= 2$ m
$l =$ length of wire $= 0.9$ m
$h = 0.4$ m

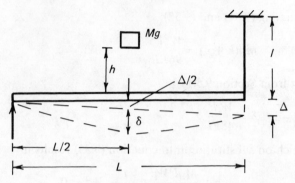

Fig. 9.17.   Impact on a beam.

9.17.2

Let,

W_e = equivalent static load to cause the deflections δ and Δ

Δ = deflection of wire, due to impact

δ = maximum central deflection of beam, due to flexure alone, on impact

i.e. maximum central deflection of beam, due to impact = $\delta + \Delta/2$.

From equilibrium considerations, the maximum force in the wire = $W_e/2$
or,

$$\Delta = \frac{W_e l}{2AE} \tag{9.55}$$

Now, from equation (9.30):

$$U_b = \frac{W_e^2 L^3}{96EI} \tag{9.56}$$

Now, from equation (9.15), the strain energy in the wire =

$$U_e = \frac{\sigma^2}{2E} * Al$$

but,

$$\sigma = E\varepsilon = \frac{E\Delta}{l} = \frac{W_e}{2A}$$

or,

$$U_e = \frac{W_e^2 * l}{8AE} \tag{9.57}$$

so that, the total strain energy =

$$U = \frac{W_e L^3}{96EI} + \frac{W_e^2 l}{8AE} \tag{9.58}$$

P.E. of mass = $Mg(h + \delta + \Delta/2)$ \hfill (9.59)

but,

$$\delta = \frac{W_e L^3}{48EI} \tag{9.60}$$

Equating (9.58) and (9.59), and by substituting equations (9.55) and (9.60), the following relationship is obtained:

$$\frac{W_e^2 L^3}{96EI} + \frac{W_e^2 l}{8AE} = Mg\left(h + \frac{W_e L^3}{48EI} + \frac{W_e l}{2AE}\right) \tag{9.61}$$

Substituting the appropriate values into equation (9.61):

$$\frac{W_e^2 \times 8}{96 \times 2E11 \times 2E\text{-}8} + \frac{W_e^2 \times 0.9}{8 \times 1.3E\text{-}6 \times 2E11}$$

$$= 2 \times 9.81\left(0.2 + \frac{W_e \times 8}{48 \times 2E11 \times 2E\text{-}8} + \frac{W_e \times 0.9}{2 \times 1.3E\text{-}6 \times 2E11}\right)$$

or,

$$W_e^2(2.083E\text{-}5 + 4.327E\text{-}7) = 3.924 + W_e(8.175E\text{-}4 + 3.396E\text{-}5)$$

therefore

$$2.127E\text{-}5 \ W_e^2 - 8.515E\text{-}4 W_e - 3.924 = 0$$

$$W_e = \frac{8.515E\text{-}4 + \sqrt{(8.515E\text{-}4)^2 + 4 \times 2.127E\text{-}5 \times 3.924}}{2 \times 2.127E\text{-}5}$$

$$= 20 + 430$$

$$\underline{W_e = 450 \text{ N}}$$

and,

Maximum stress in wire due to impact $= 346.1 \text{ MN/m}^2$

9.18.1   EXAMPLE 9.12   REINFORCED CONCRETE PILLAR UNDER IMPACT

A concrete pillar of length 4 m has a cross-sectional area of 0.25 m^2, where 10% is composed of steel reinforcement and the remainder of concrete. If a mass of 10 kg is dropped onto the top of the concrete pillar, from a height of 0.12 m above it, determine the maximum stresses in the steel and the concrete due to impact.

$$E_s = \text{elastic modulus of steel} \quad = 2 \times 10^{11} \text{ N/m}^2$$
$$E_c = \text{elastic modulus of concrete} = 1.5 \times 10^{10} \text{ N/m}^2$$
$$g = 9.81 \text{ m/s}^2$$

9.18.2

Let,

δ = the maximum deflection of the column under impact, as shown in Fig. 9.18

Work done = W.D. = $Mg(h + \delta)$ (9.62)

Strain energy = $\dfrac{\sigma_c^2}{2E_c} \times 0.225 \times 4 + \sigma_s^2 \times \dfrac{0.025 \times 4}{2E_s}$ (9.63)

where,

σ_c = maximum stress in the concrete due to impact
$\quad = E_c \varepsilon_c$
σ_s = maximum stress in the steel due to impact
$\quad = E_s \varepsilon_s$
ε_c and ε_s = maximum strain in concrete and steel, respectively

Equating (9.62) and (9.63),

$$Mg(h + \delta) = 3\text{E-}11\varepsilon_c^2 + 2.5\text{E-}13\varepsilon_s^2$$

$$98.1(0.12 + \varepsilon_c \times 4) = 3\text{E-}11\sigma_c^2 + 2.5\text{E-}13\sigma_s^2$$

$$98.1\left(0.12 + \frac{\sigma_c}{E_c} \times 4\right) = 3\text{E-}11\sigma_c^2 + 2.5\text{E-}13\sigma_s^2$$

$$11.772 + 2.616\text{E-}8\sigma_c = 3\text{E-}11\sigma_c^2 + 2.5\text{E-}13\sigma_s^2 \qquad (9.64)$$

but,

$\varepsilon_c = \varepsilon_s$

and,

$\dfrac{\sigma_c}{E_c} = \dfrac{\sigma_s}{E_s}$

or,

$$\sigma_s = \sigma_c E_s / E_c = 13.33\sigma_c \qquad (9.65)$$

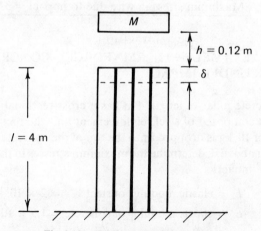

Fig. 9.18. Reinforced concrete column.

Substituting equation (9.65) into (9.64),

$$7.444E\text{-}11\sigma_c^2 - 2.616E\text{-}8\sigma_c - 11.772 = 0$$

therefore

$$\sigma_c = \frac{2.616E\text{-}8 + \sqrt{6.844E\text{-}16 + 3.505E\text{-}9}}{1.4888E\text{-}10}$$

$$\sigma_c = 0.398 \text{ MN/m}^2 \tag{9.66}$$

and from equation (9.65),

$$\sigma_s = 5.303 \text{ MN/m}^2$$

9.19.1   UNIT LOAD METHOD

This is similar to equations of the form of (9.31), and it is applicable to linear elastic structures. Using this method, the displacement at any point in a structure can be obtained from equations such as (9.67) and (9.68):

$$\delta = \int \frac{M * M_u}{EI} \, dx \tag{9.67}$$

and,

$$\delta = \int \frac{T * T_u}{GJ} \, dx \tag{9.68}$$

where M and T are the bending moment and torque, acting on an element of length "dx", due to the external loads, and M_u and T_u are the bending moment and torque acting on the same element, due to *unit loads acting in the position and direction of the required displacements.*

9.20.1   EXAMPLE 9.13   EXAMPLES BASED ON THE UNIT LOAD METHOD

Applying the unit load method to Example 9.4.
From section 9.7.1,

$$M = -WR \cos \theta - HR(1 - \sin \theta) \tag{9.69}$$

9.20.2

To determine δV, assume $W = 1$ and $H = 0$. Hence,

$$M_u = -R \cos \theta \tag{9.70}$$

Substituting (9.69) and (9.70) into (9.67),

$$\delta V = \frac{1}{EI} \int_0^{\pi/2} [WR \cos \theta + HR(1 - \sin \theta)]R \cos \theta \, R \, d\theta$$

which is identical to δV in Section 9.8.2.

9.20.3

Similarly, *if δH is required*, assume $W = 0$ and $H = 1$, so that

$$M_u = -R(1 - \sin \theta) \tag{9.71}$$

Substituting equations (9.69) and (9.71) into (9.67):

$$\delta H = \frac{1}{EI} \int_0^{\pi/2} [WR \cos \theta + HR(1 - \sin \theta)]R(1 - \sin \theta)R \, d\theta$$

which is identical to δH in Section 9.8.2.

EXAMPLES FOR PRACTICE 9

1. Determine expressions for the horizontal and vertical deflections of the free end of the thin curved beam shown in Fig. Q.9.1(a) to (c).

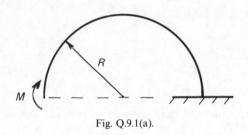

Fig. Q.9.1(a).

$\{$Horizontal deflection $= 2MR^2/EI$—to left;

Vertical deflection $= \pi MR^2/EI$—upwards$\}$

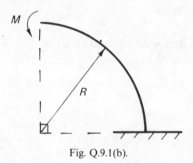

Fig. Q.9.1(b).

$\left\{\text{Horizontal deflection} = \dfrac{MR^2}{EI}\left(\dfrac{\pi}{2} - 1\right)\text{—to left;}\right.$

$\left.\text{Vertical deflection} = MR^2/EI\text{—downwards}\right\}$

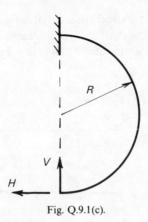

Fig. Q.9.1(c).

$$\left\{\text{Horizontal deflection} = \delta H = \frac{R^3}{EI}\left(\frac{3\pi}{2}H + 2V\right)\text{—to the left;}\right.$$

$$\left.\text{Vertical deflection} = \delta V = \frac{R^3}{EI}\left(2H + \frac{\pi V}{2}\right)\text{—upwards}\right\}$$

2. A thin curved beam consists of a length AB, which is of semi-circular form, and a length BC, which is a quadrant, as shown in Fig. Q.9.2.

Determine the vertical deflection at "A" due to a downward load W, applied at this point. The end "C" is firmly fixed, and the beam's cross-section may be assumed to be uniform.

{Portsmouth Polytechnic—March 1982}

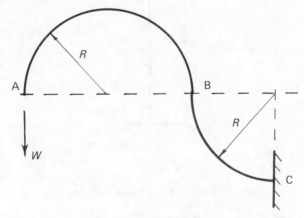

Fig. Q.9.2.

$$\left\{\delta V = \frac{WR^3}{EI}\left(\frac{25\pi}{4} - 6\right)\right\}$$

3. A clip, which is made from a length of wire of uniform section, is subjected to the load shown in Fig. Q.9.3. Determine the distance by which the free ends of the clip separate, when it is subjected to this load.

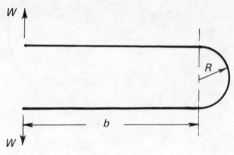

Fig. Q.9.3.

$$\left\{\frac{2W}{EI}\left[\frac{b^3}{3} + R\left(\frac{b^2\pi}{2} + \frac{R^2\pi}{4} + 2bR\right)\right]\right\}$$

4. Determine the vertical and horizontal displacements at the free end of the rigid-jointed frame shown in Fig. Q.9.4.

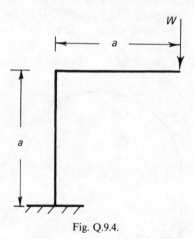

Fig. Q.9.4.

$$\{\delta V = 4Wa^3/(3EI)\text{—downwards};$$

$$\delta H = Wa^3/(2EI)\text{—to the right}\}$$

5. Determine the horizontal deflection at the free end "A" of the rigid-jointed frame of Fig. Q.9.5.

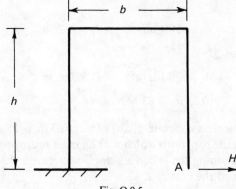

Fig. Q.9.5.

$$\left\{ \delta H = \frac{Hh^2}{EI} \left(\frac{2h}{3} + b \right) \right\}$$

6. If the framework of Fig. Q.9.5 were prevented from moving vertically at "A", so that it was statically indeterminate to the first degree and $h = b$, what force would be required to prevent this movement?

$\{0.75H,\ \text{acting downwards}\}$

7. A rod, composed of two elements of different cross-sectional areas, is firmly fixed at its top end, and has an inextensible collar, firmly secured to its bottom end, as shown in Fig. Q.9.7.

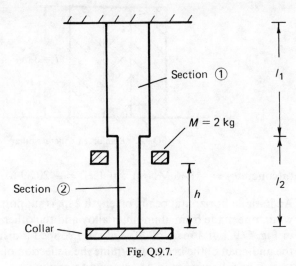

Fig. Q.9.7.

If a mass of magnitude 2 kg is dropped from a height of 0.4 m above the top of the collar, determine the maximum deflection of the rod, and also the maximum stress.

$$E = 2\mathrm{E}11\ \mathrm{N/m^2}, \qquad g = 9.81\ \mathrm{m/s^2}$$

Section ①

$$A_1 = 500 \text{ mm}^2; \qquad l_1 = 1.2 \text{ m}$$

Section ②

$$A_2 = 300 \text{ mm}^2; \qquad l_2 = 0.8 \text{ m}$$

$$\{0.63 \text{ mm}; 83 \text{ MN/m}^2\}$$

8. A reinforced concrete pillar, of length 3 m, is fixed firmly at its base and is free at the top, onto which a 20 kg mass is dropped, as shown in Fig. Q.9.8.

   Assuming the following apply, determine the maximum stresses in the steel and the concrete:

Steel

$$A_s = \text{cross-sectional area of steel reinforcement} = 0.01 \text{ m}^2$$
$$E_s = \text{elastic modulus in steel} = 2\text{E}11 \text{ N/m}^2$$

Concrete

$$A_c = \text{cross-sectional area of concrete reinforcement} = 0.2 \text{ m}^2$$
$$E_c = \text{elastic modulus in concrete} = 1.4\text{E}10 \text{ N/m}^2$$

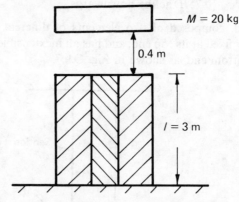

Fig. Q.9.8. Reinforced concrete pillar.

$$\{\sigma(\text{concrete}) = -1.46 \text{ MN/m}^2; \sigma(\text{steel}) = -20.89 \text{ MN/m}^2\}$$

9. An initially horizontal beam, of length 2 m, is supported at its ends by two wires, one made from aluminium alloy and the other from steel, as shown in Fig. Q.9.9. If a mass of 203.87 kg is dropped a distance of 0.1 m above the mid-span of the beam, determine the deflection of the beam, under this load. The following may be assumed to apply:

Steel

$$E_s = 2\text{E}11 \text{ N/m}^2$$
$$l_s = 2 \text{ m}$$
$$A_s = 2\text{E-4 m}^2$$

Aluminium Alloy

$E_A = 7 \times 10^{10}$ N/m^2

$l_A = 1$ m

$A_A = $ 4E-4 m^2

EI (for beam) = 2E6 N/m^2

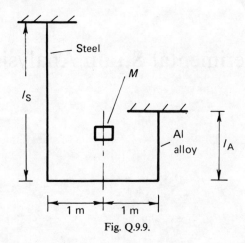

Fig. Q.9.9.

$\{\delta = 6.49$ mm; $\sigma_s = 154.98$ MN/m^2; $\sigma_{Al} = 77.48$ MN/m$^2\}$

10

Experimental Strain Analysis

10.1.1

In this chapter, a brief description will be given of some of the major methods in experimental strain analysis, and in particular, to the use of electrical resistance strain gauges and photoelasticity.

The aim in this chapter is to expose the reader to various methods of experimental strain analysis, and to encourage him/her to consult other publications, which cover this topic in a more comprehensive manner [23–25].

10.2.1   ELECTRICAL RESISTANCE STRAIN GAUGES

0.1%
= 1000 µ strain

The elastic strain in most structures, constructed from steel, aluminium alloy, etc., seldom exceeds 0.1%, and it is evident that such small magnitudes of strain will need considerable magnification to record them precisely. This feature presented a major problem to structural engineers in the past, and it was not until the 1930s that this problem was resolved, when the electrical resistance strain gauge was invented. This gauge was invented as a direct result of requiring lighter aircraft structures, although the principle that the electrical resistance strain gauge is based on was discovered as early as 1856, by Lord Kelvin, when he observed that the electrical resistance of copper and iron wires varied with strain. The discovery of Lord Kelvin was that, when a length of wire is strained, its electrical resistance changes and, within certain limits, this relationship is linear, and can be expressed as follows:

$$\text{Strain} \propto \text{change of electrical resistance; or strain } (\varepsilon) = K\frac{\Delta R}{R}$$

$$(10.1)$$

where,

> K is known as the gauge factor, and it is dependent on the material of construction (i.e. it is a material constant). For an ordinary Cu/Ni gauge, K is about 2.
>
> R is the electrical resistance of the strain gauge (ohms).
>
> ΔR is the change of electrical resistance of the strain gauge (ohms) due to ε.

As K is known, and R and ΔR can be measured, ε can be readily obtained from equation (10.1).

10.2.2   Temperature Compensation

If a strain gauge is subjected to a change of temperature, it very often suffers a larger change in length than that caused by external loadings. To overcome this deficiency, it is necessary to attach another strain gauge (called a "dummy' gauge) to a piece of material with the same properties as the structure itself, and to subject this piece of material to the same temperature changes as the structure, but not to "constrain" the "dummy" gauge, or allow it to undergo any external loading.

The dummy gauge should have identical properties as the active gauge, and for static analysis, the two gauges should be connected together in the form of a Wheatstone bridge, as shown in Fig. 10.1.

10.2.3   Pressure Compensation

If an electrical resistance strain gauge is subjected to fluid pressure, its electrical resistance changes owing to the Poisson effect of the gauge material. To overcome this deficiency, it is necessary to subject the strain gauge to the same pressure medium conditions as the active gauge.

10.2.4   The "Null" Method of Measuring Strains

Fig. 10.1 shows the circuitry for this method of strain measurement, which is suitable for static analysis.

Although the active gauge and the dummy gauge may have the same initial electrical resistance, after attaching these gauges to their respective surfaces, there may be small differences in their electrical resistances, so that the galvanometer will become unbalanced. Thus, prior to loading the structure, it will be necessary to balance the galvanometer, by suitably changing the electrical resistance on the variable resistance arm. Once the structure is loaded, and the gauge experiences strain, the galvanometer will once again become unbalanced, and by balancing the galvanometer, the change of electrical resistance due to this load can be measured. Normally, the strain gauge equipment has facilities which allow it to directly record strain, providing the gauge factor is pre-set, for the particular batch of gauges.

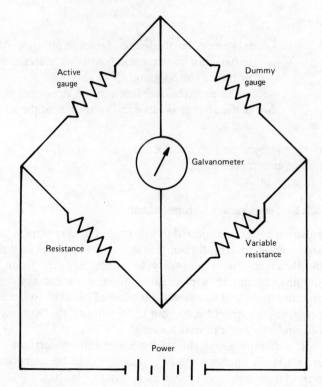

Fig. 10.1.   The "null" method of measuring strains.

Further loading of the structure will cause the galvanometer to become unbalanced, once again, and by rebalancing the galvanometer, the strain can be recorded for this particular loading condition.

10.2.5   The "Deflection" Method of Measuring Strain

It is evident from Section 10.2.4, that the "null" method is only suitable for static analysis, where there is a sufficient time available to record each individual strain. Thus, for dynamic analysis, or where many measurements of static strain are required, in quick succession, the "null" method is unsuitable.

One method of overcoming this problem is to use the "deflection" method of strain analysis, where the strain gauge circuit of Fig. 10.2 is used.

In this case, as the measured strains are very small, they need to be amplified prior to being recorded. The strain recorder can take many forms, varying from chart recorders and storage oscilloscopes to computers. Thus, it is evident that this method of measuring strain is much more expensive than the null method, because of its requirements for large-scale amplification and sensitive strain recorders, but, nevertheless, it has been successfully employed for dynamic strain recording. The electrical supply shown in Fig. 10.2 can be either A.C. or D.C.

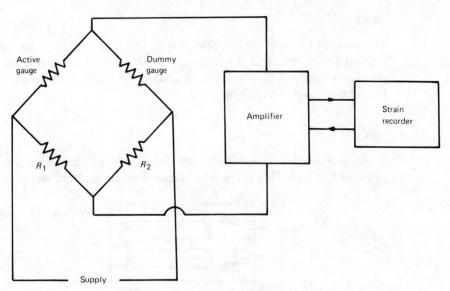

Fig. 10.2.   Strain gauge circuit for the "deflection" method of strain measurement.

10.3.1   Types of Electrical Resistance Strain Gauge

The two most common types of electrical resistance strain gauge are the wire gauge and the foil gauge, and these gauges are now described.

10.3.2   Wire Gauges

One of the simplest forms of the wire gauge is the *zigzag* gauge, where the diameter of its wire can be as small as 0.025 mm, as shown in Fig. 10.3. The reason why this gauge wire is of zigzag form is because a minimum length of wire is required, so that the supplied power can be sufficiently small to prevent heating of the gauge itself. The gauge is constructed by sandwiching a length of high resistance wire in zigzag form, between two pieces of paper.

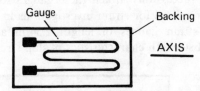

Fig. 10.3.   A linear zigzag wire gauge.

It should be noted that the gauge is attached to the structure via its backing, and that the latter is required as an electrical insulation. The problem with the zigzag gauge is that it is prone to cross-sensitivity.

10.3.3

Cross-sensitivity is an undesirable property of a strain gauge, as the high resistance wire, which is perpendicular to the axis of the gauge, measures erroneous strains in this direction, in addition to the required strains, which lie along the axis of the gauge.

10.3.4

To some extent, cross-sensitivity can be reduced by employing the strain gauge of Fig. 10.4. This gauge is constructed by placing thin strands of high

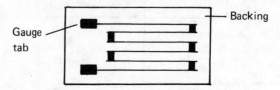

Fig. 10.4.   A linear wire gauge, with welded cross-bars.

resistance wire parallel to each other, and connecting them together by low resistance welded cross-bars. Thus, as the electrical resistance of the cross-bars is small compared with the electrical resistance of the wires, the effects of cross-sensitivity are much smaller than they are for zigzag wire gauges.

The backing material of these gauges can be either of paper or of plastic.

10.3.5   Foil Gauges

Another popular strain gauge is the foil gauge of Fig. 10.5, which was invented by Saunders-Roe in 1952. This gauge is etched out of flat metal foil, so that its electrical resistance along its axis can be very large compared with its electrical resistance perpendicular to its axis. This property allows successful construction of such gauges to be as small as 0.2 mm (1/128 inch) or as long as 2.54 m (100 inches), with negligible cross-sensitivity effects. In the cases of these extreme lengths of strain gauge, considerable expertise is required to successfully attach these gauges to the structure. Small strain gauges are required for strain investigations in regions of stress concentrations, and strain gauges of length 2.54 m are used to investigate the longitudinal strength characteristics of ship structures.

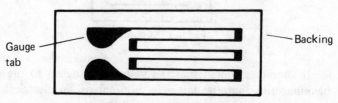

Fig. 10.5.   A linear foil gauge.

In general, the most popular lengths of linear gauges vary between about 5 and 10 mm. The main advantage of the foil gauge is that because it is etched out of flat metal foil, it and its backing can be very thin, so that the gauge can be in a more intimate contact with the structure, than can a wire gauge.

Disadvantages of foil gauges are that they are generally more expensive than wire gauges, and that for very thin foil gauges, they can easily fracture, if roughly handled. Foil gauges are usually plastic backed.

10.4.1   Gauge Material

Most electrical resistance strain gauges of normal resistance (about 100 Ω) are constructed from a copper–nickel alloy, where the proportion of Cu : Ni usually varies from about 55 : 45 to 60 : 40. For such gauges, the gauge factor is about 2, and depends on the Cu : Ni ratio.

A popular alloy for high resistance gauges (about 1000 Ω) is called NICHROME, which consists of 75% Ni, 12% Fe, 11% Cr and 2% Mn.

10.5.1   Gauge Adhesives

In the electrical resistance strain gauge technique, this is one of the most important considerations. It is important that the experimentalist uses a satisfactory adhesive, as an unsatisfactory adhesive can cause hysteresis or zero drift. In addition to this, another important consideration is the preparation of the surface that the gauge is to be attached to.

Prior to attaching the gauge, it should be ensured that the surface it is to be attached to is clean and free from rust, paint, dirt, grease, etc. This surface should not be highly polished; if it is, it should be roughened a little with a suitable abrasive, and then it should be degreased with a suitable solvent, such as acetone, alcohol, etc. After application of the strain gauge cement to the appropriate part of the structure's surface, and in some cases to the back of the gauge itself, the gauge should be gently placed onto its required position.

To ensure that no air bubbles become trapped between the gauge and the surface of the structure, the thumb, or finger, should be pressed firmly onto the surface of the gauge, and gently rolled to and fro, until the excess strain gauge cement is squeezed out.

The adhesion time of the gauge varies from a few minutes to several days, depending on the type of adhesive used and the environment that the adhesive is to be exposed to.

Gauge adhesives are generally either organic based or ceramic based, the former being satisfactory for temperatures below 260 C and the latter for temperatures in excess of this. A brief description of some of the different types of gauge adhesive will now be given.

10.5.2   Cellulose Acetates

These are among the most common forms of adhesive (e.g. Durofix) that can be bought from high street shops. They adhere by evaporation, and take from 24 hours to 3 days to gain full strength, depending on the surrounding temperature. They are usually used for paper-backed gauges.

10.5.3   Epoxy Resins

These require the mixing of a resin with a hardener. A popular combination is to use Araldite strain gauge cement, with either Araldite hardener HY951 or Araldite hardener HY956, in the following proportions by weight:

	Araldite strain gauge cement	100
	with either HY951 hardener	4 to 4.5
or	HY956 hardener	8 to 10

Another epoxy resin adhesive is the M-Bond epoxy supplied by Welwyn Strain Measurements.

These cements adhere by curing, where the time taken to gain full strength can take 24 hours, but in the case of Araldite, the curing time can be accelerated by exposing the cement to ultra-violet radiation. These adhesives are suitable for either plastic-backed or paper-backed gauges.

10.5.4   Cyanoacrylates

These are pressure adhesives, which adhere by applying pressure to the cement, via the gauge. One of the earliest of these adhesives is called Eastman-Kodak 910, which allows a strain gauge to be used within a few minutes of attaching it to the structure.

Another more recent cyanoacrylate is M-Bond 200, which is supplied by Welwyn Strain Measurements.

10.5.5   Norton-Rokide

This adhesive is suitable for high temperature work, and because of this, the strain gauge does not normally have any backing. Instead, the cement is first sprayed onto the surface, and then the gauge is firmly pressed down, but care has to be taken to ensure that some cement lies between the gauge and the structure, so that its resistance to earth does not break down. After the wires have been attached to the gauge tabs, further cement is sprayed over the gauge and the surrounding surface, so that it is encapsulated.

10.6.1   Water-proofing

If a strain gauge is exposed to water or damp conditions, its electrical resistance to earth will break down, and render the gauge useless. Thus, if a gauge is likely to be exposed to such an environment, it is advisable to waterproof the gauge and its wiring.

Some methods of water-proofing strain gauges are discussed below, but prior to using any of these methods, it should be assumed that the gauges and their surrounding surfaces are free from water.

10.6.2   Di-Jell

This is a micro-crystalline wax, which has the appearance of a jelly-like substance. In general, it is only suitable for damp-proofing or for water-proofing when the water is stationary. To water-proof the gauge, the Di-Jell is simply applied to the gauge and its surrounding surface.

10.6.3

Silicone greases and petroleum jelly (Vaseline) can also be used for damp-proofing, but care should be taken to ensure that these substances are not subjected to temperatures which will melt them.

10.6.4

A more robust and permanent method of water-proofing strain gauges is that recommended by Welwyn Strain Measurement Limited. This consists of painting M-Coat A or D over the gauge and its adhesive, and then covering the M-Coat with aluminium foil. Finally, the whole surface is covered with M-Coat G, the electrical leads being first covered with M-Coat B, as shown in Fig. 10.6.

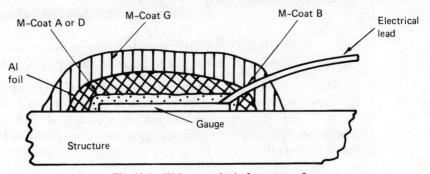

Fig. 10.6.   Welwyn method of water-proofing.

Prior to the development of the Welwyn method of water-proofing strain gauges, the Saunders-Roe technique proved popular. This method consisted of applying successive coats of epoxy resin and glass cloth, and covering the whole with an impervious rubber-based solution.

10.6.5

Other methods of water-proofing consist of simply covering the gauge and its surrounding surface with various types of sealant, including automobile underbody sealants.

Table 10.1. Effect of pressure and water immersion on the electrical resistance of strain gauges

	Electrical resistance (Ω)				
Sealant	Before	1 hour in water	3 weeks in water	3.45 MPa	6.9 MPa
Bostik 6	55	54	55	55	55
Underbody seal	56	56	55	56	56
M-Coat A	56	56	55	56	56
M-Coat D	56	55	55	55	55
Di-Jell	52	52	52	—	—

Line [26] carried out an investigation on the water-proofing qualities of a number of sealants, as shown in Table 10.1. Line used foil gauges of dimensions 2.54 cm length, 1.02 cm width. He measured the electrical resistance of these strain gauges in air, before immersion into water, and then he took these measurements again, one hour and also three weeks later, after continuous immersion in water. He also measured the electrical resistances of these gauges at water pressures of 3.45 and 6.9 MPa. He found that all the sealants were satisfactory under test, and he recorded the following observations:

(a) The underbody seal appeared to have softened, and although it was still water-proof, it could easily be chipped off by hand.
(b) Under the M-Coat A, it was possible to see that the metal was completely rust-free.
(c) The pressures were to low and the instruments were too insensitive to record a change of resistance due to the effects of pressure.

N.B. Dally and Riley [25] report of tests by Milligan [27], and by Brace [28], who found that the effect of pressure caused a strain of about 0.58×10^{-6} per MPa of pressure, and they concluded that for most problems, the effects of pressure on strain gauges can be ignored at pressures below 20.7 MPa (3000 lbf/in^2).

10.6.6

Other forms of gauge include *shear pairs* and *strain rosettes*, as described in Chapter 7, together with *crack measuring* and *diaphragm* gauges. Diaphragm gauges consist of a combination of radial and circumferential gauges, and crack measuring gauges consist of several parallel strands of wire and tabs.

10.7.1 GAUGE CIRCUITS

Skilful use of strain gauge circuits can eliminate the use of dummy gauges and provide increased sensitivity.

10.7.2   Combined Bending and Torsion of Circular Section Shafts

For circular section shafts, under the effects of combined bending and torsion, it is convenient to use two pairs of "shear pairs", as shown in Fig. 10.7. By fitting two pairs of shear pairs, it is possible to record either the bending moment M or the torque T, depending on the circuit used.

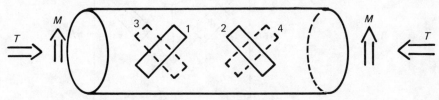

Fig. 10.7.   Shaft under combined bending and torque.

The shear pairs must be fitted at 45° to the axis of the shaft, and a typical shear pair of strain gauges is shown in Fig. 10.8. *Shear pairs* are so-called, because under pure torque, the maximum principal stresses in a circular section shaft, which are numerically equal to the maximum shear stress, lie at 45° to the axis of the shaft (see Chapter 7).

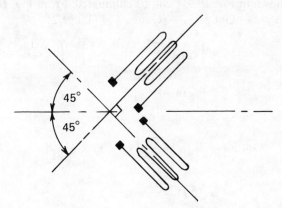

Fig. 10.8.   A shear pair of strain gauges.

10.7.3   To Determine M

Let,

ε_{45B} = direct strain due to M, which lies at 45° to the axis of the shaft

ε_{45T} = direct strain due to T, which lies at 45° to the axis of the shaft

= $\gamma/2$ (see Chapter 7)

γ = maximum shear strain, due to T

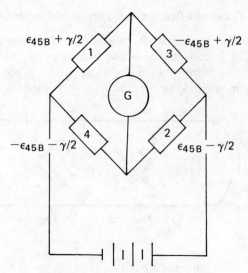

Fig. 10.9.   Circuit for measuring M.

If the gauges are connected together in the form of a full Wheatstone bridge, as shown in Fig. 10.9, γ will be eliminated. From Fig. 10.9, the *output* will be (Gauge 1 − Gauge 3) − (Gauge 4 − Gauge 2)

$$= \left(\varepsilon_{45B} + \frac{\gamma}{2} + \varepsilon_{45B} - \frac{\gamma}{2} \right) - \left(- \varepsilon_{45B} - \frac{\gamma}{2} - \varepsilon_{45B} + \frac{\gamma}{2} \right)$$

$$= 4\varepsilon_{45B}$$

or,

$$\text{Output} = 4\varepsilon_{45B} = \frac{\sigma_{45B}}{E}(1 - v) \times 4$$

$$= \frac{\sigma_B}{2E}(1 - v) \times 4$$

$$= 2\sigma_B(1 - v)/E$$

therefore

$$\sigma_B = \frac{\text{Output}}{2(1 - v)} * E = \text{bending stress}$$

but,

$$M = \sigma_B * \frac{\pi d^4}{64} * \frac{2}{d}$$

$$\underline{M = \frac{\pi d^3 \sigma_B}{32}} \tag{10.2}$$

10.7.4　To Determine T

By adopting the circuit of Fig. 10.10, ε_{45B} can be eliminated. The *output* from the circuit of Fig. 10.10

$$= (\text{Gauge 1} - \text{Gauge 2}) - (\text{Gauge 4} - \text{Gauge 3})$$

$$= \left(\varepsilon_{45B} + \frac{\gamma}{2} - \varepsilon_{45B} + \frac{\gamma}{2} \right) - \left(-\varepsilon_{45B} - \frac{\gamma}{2} + \varepsilon_{45B} - \frac{\gamma}{2} \right)$$

Output $= 2\gamma$

or,

$$\gamma = \text{Output}/2$$

but,

$$T = \tau * \frac{\pi d^4}{32} * \frac{2}{d}$$

or,

$$T = \frac{\pi G \gamma d^3}{16} \tag{10.3}$$

N.B.　Strains due to axial loads, including *thermal effects*, will automatically be eliminated when the circuits of Figs. 10.9 and 10.10 are adopted.

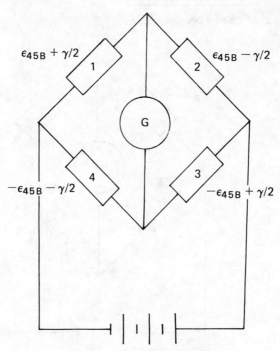

Fig. 10.10.　Circuit for measuring T.

10.7.5   Combined Bending and Axial Strains

Consider a length of beam under combined bending and axial load, as shown in Fig. 10.11.

Let,

ε_B = bending strain due to M

ε_D = direct strain due to P, plus thermal effects

The bending moment M can be obtained by adopting the circuit of Fig. 10.12.

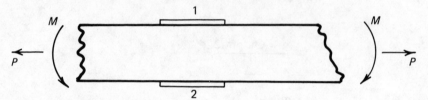

Fig. 10.11.   Combined M and P.

The output from the circuit of Fig. 10.12

$$= \text{Gauge 1} - \text{Gauge 2}$$

$$= \underline{2\varepsilon_B}$$

or,

$$\underline{\varepsilon_B = \text{Output}/2 = \sigma_B/E}$$

Hence, M can be obtained.

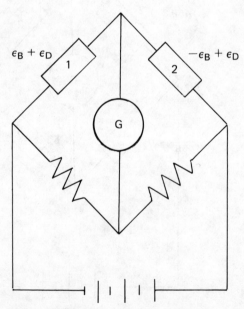

Fig. 10.12.   Circuit to determine M.

10.7.6   Full Bridge for Measuring M

$$\text{Output} = (\text{Gauge 1} - \text{Gauge 2}) - (\text{Gauge 4} - \text{Gauge 3})$$

$$= \underline{4\varepsilon_B}$$

i.e. the circuit of Fig. 10.13 will give four times the sensitivity of a single strain gauge, and twice the sensitivity of the circuit of Fig. 10.12. It will also automatically eliminate thermal strains.

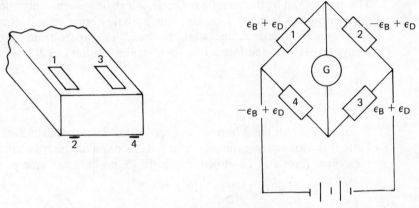

Fig. 10.13.   Full bridge for determining M.

10.7.7   Half-bridge for Measuring Axial Strains (see Fig. 10.14)

$$\text{Output} = 2\varepsilon_D = 2 * \sigma_D/E$$

Hence, P can be obtained.

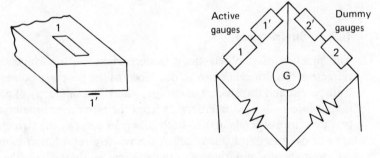

Fig. 10.14.   Circuit for eliminating bending and thermal strains.

10.8.1   PHOTOELASTICITY

The practice of photoelasticity is dependent on the shining of light through transparent or translucent materials, but prior to discussing the method, it will be necessary to make some definitions.

A beam of ordinary light vibrates in many planes, transverse to its axis of propagation. It is evident that, as light vibrates in so many different planes, it will be necessary for a successful photoelastic analysis, to only allow those components of a beam of light to vibrate in a known plane.

10.8.2

This is achieved by the use of a *polariser*, which, in general, only allows those components of light to pass through it that vibrate in a vertical plane. Polarisers are made from polaroid sheet, and apart from photoelastic analysis, they are used for sunglasses ("shades" in the U.S.A.).

10.8.3

A polariser which has a horizontal axis of transmission is called an *analyser*. Thus, if the axes of transmission of a polariser and an analyser are at 90° to each other (crossed), as shown in Fig. 10.15, no light will emerge from the analyser.

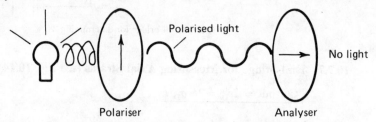

Fig. 10.15. Polariser and an analyser.

10.8.4 Birefringence

The practice of photoelasticity is dependent on birefringence or double refraction. Birefringence can be described as the property possessed by some transparent and translucent materials, whereby a single ray of polarised light is split into two rays, on emerging from the birefringent material, where the two rays are perpendicular to each other. In general, the two emerging rays are out-of-phase, and this is called the relative retardation. Some materials are permanently birefringent; others, such as that which the models are constructed from, are only birefringent under stress. In the case of the latter, a single ray of polarised light is split into two rays, on emerging from the model, where the directions of the two rays lie along the planes of the principal stresses. The relative retardation of the two rays, namely R_t, is proportional to the magnitude of the principal stresses and the thickness of the model, as follows:

$$R_t = C(\sigma_1 - \sigma_2)h \qquad (10.4)$$

where

C = the stress optical coefficient (i.e. it is a material constant)
σ_1 = maximum principal stress
σ_2 = minimum principal stress
h = model thickness

Typical materials used for photoelastic models include:

epoxy resin (e.g. Araldite), polycarbonate (e.g. Lexan or Makrolan), urethane rubber, etc.

10.8.5

A Plane Polariscope is one of the simplest pieces of equipment used for photoelasticity. It consists of a light source, a polariser, a model, an analyser and a screen, as shown in Fig. 10.16.

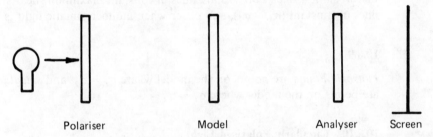

Polariser   Model   Analyser   Screen

Fig. 10.16.   A plane polariscope.

When the model is unstrained, no light will emerge, but when the model is strained, strain patterns will appear on the screen. This can be explained, as follows.

The polarised light emerges from the model along the directions of the planes of the principal stresses, so that these rays will vibrate at angles to the vertical plane. On emerging from the analyser, only those components of light which vibrate in a horizontal plane will be displayed on the screen, to give a measure of the stress distribution in the model. If daylight or white light is used in a plane polariscope, the stress patterns will involve all the colours of the rainbow. The reason for this is that daylight or white light is composed of all the colours of the rainbow, each colour vibrating at a different frequency. The approximate frequency of deep red light is about 390×10^{12} Hz, and the approximate frequency of deep violet light is 770×10^{12} Hz, the speed of light in a vacuum being about 2.998×10^8 m/s (186 282 miles/s). The difference in frequencies between different coloured lights is one of the reasons why violet is on the inside of the rainbow and red is on the outside, and these differences probably account for why we can distinguish between different colours.

However, whereas the stress patterns are quite spectacular when daylight is used, it is difficult to analyse the model. For this reason, it is preferable to use *monochromatic light*.

10.8.6

Monochromatic Light is light of one wavelength only, and when it is used, the stress patterns consist of dark lines against a light background. Typical lamps used to produce monochromatic light include mercury vapour and sodium.

The use of monochromatic light in a plane polariscope will produce *isoclinics* and *isochromatics*, which are defined as follows.

10.8.7

Isoclinic Fringe Patterns are lines of constant stress direction, and occur when monochromatic light is used. They are useful for determining the principal stress directions.

10.8.8

Isochromatic Fringe Patterns are lines of constant maximum shear stress, i.e. lines of constant $(\sigma_1 - \sigma_2)$, and occur when monochromatic light is used.

10.8.9

Isotropic Points are points on the model where $\sigma_1 = \sigma_2$, and *singular* points are points on the model where $\sigma_1 = \sigma_2 = 0$.

10.8.10   Circularly Polarised Light

One of the problems with using a plane polariscope is that both isochromatic and isoclinic fringes appear together, and much difficulty is experienced in distinguishing between the two. The problem can be overcome by using a circular polariscope, which can extinguish the isoclinics. A plane polariscope can be converted to a circular polariscope by inserting quarter-wave plates, as shown in Fig. 10.17.

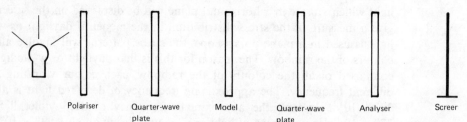

Polariser   Quarter-wave   Model   Quarter-wave   Analyser   Screen
                   plate   plate

Fig. 10.17.   Circular polariscope.

10.8.11

Quarter-wave Plates are constructed from permanently birefringent materials, which have a relative retardation of $\lambda/4$, and to extinguish the isoclinics, they are placed with their fast axes at 45° to the planes of polarisation of the polariser and the analyser, and at 90° to each other, were,

λ = the wavelength of the selected light

10.8.12 Some Notes on Experimental Photoelasticity

To obtain the isoclinics, remove the quarter-wave plates, and rotate the polariser and the analyser by the same angle and in the same direction. This process will make the isoclinic fringes move with the change of angle, but will not affect the isochromatic fringes.

To obtain *isochromatic half-fringes* in a circular polariscope, place the axes of polarisation of the polariser in the same plane as the axis of polarisation of the analyser.

The *isochromatic fractional fringes* can be obtained by appropriately placing the plane of the polariser at various angles, other than 0° and 90°, to the plane of polarisation of the analyser.

10.8.13 Material Fringe Value f

The material fringe value for any given photoelastic material is given by

$$f = \frac{\lambda}{2C} \ N/(\text{m.fringe}) \tag{10.5}$$

where,

C = stress optical coefficient
λ = wavelength of the selected light

If there are n *isochromatic fringes*, then from equations (10.4) and (10.5),

$$\frac{\sigma_1 - \sigma_2}{2} = \frac{n * f}{h} = \hat{t}$$

or,

$$\hat{t} = n * f/h \tag{10.6}$$

where,

h = model thickness
$\hat{t}$ = maximum shear stress at the nth fringe.

Thus, at the ith fringe, the maximum shear stress τ_i is given by

$$\hat{t}_i = i * f/h$$

Typical values of f, under mercury green light, together with other material constants for some photoelastic materials are given in Table 10.2.

Table 10.2. Some material constants for photoelastic materials

Material	f (kN/m.fringe)	E (MPa)	v	U.T.S.* (MPa)
Araldite B	10.8	2760	0.38	70
Makralon	7.0	2930	0.38	62
Urethane rubber	0.2	3	0.46	0.15*

* This quoted value is not the ultimate tensile strength (U.T.S.) of the material, but its elastic limit.

10.8.14 Stress Trajectories

These are a family of lines which lie orthogonally to each other. They are lines of constant principal stress, and they can be constructed from isoclinics.

10.9.1 Three-dimensional Photoelasticity

One popular method in three-dimensional photoelasticity is to load the model whilst it is being subjected to a temperature between 120 and 180°C. The model is then allowed to cool slowly to room temperature, but the load is not removed during this process. On removing the load, when the model is at room temperature, the stress system will be "frozen" into the model, and then by taking thin two-dimensional slices from the model, a photoelastic investigation can be carried out—sometimes by immersing the slides into a liquid of the same refractive index as the model's material.

There are many other methods in three-dimensional photoelasticity, but descriptions of these methods are beyond the scope of this book.

10.10.1 MOIRÉ FRINGES

These have no connection whatsoever with the photoelastic method. They are in fact interference patterns whereby a suitable pattern is placed or shone onto the structure. The pattern is then noted or photographed before and after deformation, and the two patterns are superimposed to produce "fringes". Examination of these fringes by the use of a comparator or a microscope can be made to determine the experimental strains.

The patterns can consist of lines or grids or dots, etc., which can be parallel, radial or concentric, depending on the shape of the structure to be analysed.

10.11.1 BRITTLE LACQUER TECHNIQUES

This is another method of experimental strain analysis, where a thin coating of a brittle lacquer is sprayed or painted onto the surface of the structure, before it is loaded. After loading the structure, the lacquer will be found to have cracked patterns, and these patterns will be related to the direction and magnitude of the maximum tensile stresses. One of the most popular brittle lacquers is called "stress-coat", which is manufactured by the Magnaflux Corporation of the U.S.A.

One of the problems with using brittle lacquer is that the cracks only occur under tension, so that if regions of compressive stresses are required to be examined, the lacquer must be applied to the surface of the structure while it is under load. On removing the load, the brittle lacquer will crack in those zones where the structure was in compression when loaded.

10.12.1   SEMI-CONDUCTOR STRAIN GAUGES

These are constructed from a single crystal of silicon, and each gauge takes the form of a short rectangular filament.

Semi-conductor strain gauges are very sensitive to strain, where their change of resistance for a given strain can be about 100 times the value of the change of resistance of an electrical resistance strain gauge.

Semi-conductor strain gauges are usually small (e.g. 0.5 mm), and because of this and their high sensitivity, they are particularly useful for experimental strain analysis in regions of high stress concentration. Their main disadvantage is that they are much more expensive than electrical resistance strain gauges. They are usually manufactured from either a P-type material or an N-type material. The gauges can be made to be self-temperature compensating, by constructing the gauge from a P-type element together with an N-type element—the two being connected in one half of a Wheatstone bridge.

10.13.1   ACOUSTICAL GAUGES

These gauges are based on measuring the magnitude of the resonating frequency of a piece of wire, stretched between two knife edges. Initially, the wire must be in tension, so that any compressive strains it is likely to receive under load will not cause it to lose its bending stiffness. The wire is "plucked" by an electromagnet, and another electromagnet, called the "pick-up", receives the signals from the resonating wire. The signal received by the pick-up magnet is then amplified, and apart from the signal being sent to an oscilloscope, it is used to further excite the "plucking" magnet, so that the amplitude of the resonating wire will be maximised.

One knife edge of the acoustical gauge is fixed, and the other knife edge is movable, so that the latter can transmit strain to the resonating wire. Change of length of the wire will cause its resonant frequency to change, which can then be compared with a reference acoustical gauge.

The reference acoustical gauge can then be adjusted with the aid of a micrometer screw gauge attached to it, so that its resonant frequency is the same as the gauge under test, and, hence, the strain recorded.

Acoustical gauges vary in length from 2.54 cm (1 inch) to 15.24 cm (6 inches), and because of their bulk, they are generally only preferred to electrical resistance strain gauges in special circumstances, where their inherent robustness and long-term stability characteristics are considered to be of prime importance.

References

[1] Ross, C. T. F. *Computational Methods in Structural and Continuum Mechanics*, Horwood, 1982.
[2] Williams, J. G. *Stress Analysis of Polymers*, Longman, 1973.
[3] Macaulay, W. H. Note on the Deflection of Beams, *Messenger of Mathematics*, **48**, 129–130, 1919.
[4] Stephens, R. C. *Strength of Materials*, Arnold, 1970.
[5] Ross, C. T. F. *Finite Element Methods in Structural Mechanics*, Horwood, 1985.
[6] Johnson, D. *Advanced Structural Mechanics*, Collins, 1986.
[7] Rockey, K. C., Evans, H. R., Griffiths, D. W. and Nethercot, D. A. *The Finite Element Method*, 2nd Edition, Collins, 1983.
[8] Saada, A. S. *Elasticity—Theory and Applications*, Pergamon, 1974.
[9] Den Hartog, J. P. *Advanced Strength of Materials*, McGraw-Hill, 1952.
[10] Timoshenko, S. P. and Goodier, J. N. *Theory of Elasticity*, McGraw-Hill-Kogakusha, 1970.
[11] Ford, Hugh and Alexander, J. M. *Advanced Strength of Materials*, Horwood, 1977.
[12] Ross, C. T. F. *Finite Element Programs for Axisymmetric Problems in Engineering*, Horwood, 1984.
[13] Ross, C. T. F. The Collapse of Ring-reinforced Circular Cylinders under Uniform External Pressure, *Trans. R.I.N.A.*, **107**, 375–394, 1965.
[14] Richards, T. H. *Energy Methods in Stress Analysis*, Horwood, 1977.
[15] Zienkiewicz, O. C. *The Finite Element Method*, 3rd Edition, McGraw-Hill, 1977.
[16] Cook, R. D. *Concepts and Applications of Finite Element Analysis*, 2nd Edition, Wiley, 1981.
[17] Segerlind, L. J. *Applied Finite Element Analysis*, 2nd Edition, Wiley, 1984.

[18] Irons, B. and Ahmad, S. *Techniques of Finite Elements*, Horwood, 1980.

[19] Fenner, R. T. *Finite Element Methods for Engineers*, Macmillan, 1975.

[20] Davies, G. A. O. *Virtual Work in Structural Analysis*, Wiley, 1982.

[21] Smith, I. M. *Programming the Finite Element Method*, Wiley, 1982.

[22] Owen, D. R. J. and Hinton, E. *A Simple Guide to Finite Elements*, Pineridge, Swansea, 1980.

[23] Coker, E. G. and Filon, L. N. G. *A Treatise on Photoelasticity*, Cambridge University Press, 1931.

[24] Holister, G. S. *Experimental Stress Analysis—Principles and Methods*, Cambridge University Press, 1967.

[25] Dally, J. W. and Riley, W. F. *Experimental Stress Analysis*, 2nd Edition, McGraw-Hill-Kogakusha, 1978.

[26] Line, D. R. Investigation into the Use of Some Commercial and Industrial Adhesives and Sealants for Electrical Resistance Strain Gauge Application, Final Year Project, Dept. of Mech. Eng., Portsmouth Polytechnic, 1977/78.

[27] Milligan, R. V. The Effects of High Pressure on Foil Strain Gages, *Expt. Mech.*, **4**, 25–36, 1964.

[28] Brace, W. F. Effect of Pressure on Electrical Resistance Strain Gages, *Expt. Mech.*, **4**, 212–216, 1964.

Index